THE GOLD CRUSADES:
A SOCIAL HISTORY OF GOLD RUSHES, 1849–1929

REVISED EDITION

Among the hordes of starry-eyed 'argonauts' who flocked to the California gold rush of 1849 was an Australian named Edward Hargraves. He left America empty-handed, only to find gold in his own backyard. The result was the great Australian rush of the 1850s, which also attracted participants from around the world. A South African named P.J. Marais was one of them. Marais too returned home in defeat – only to set in motion the diamond and gold rushes that transformed southern Africa. And so it went.

Most previous historians of the gold rushes have tended to view them as acts of spontaneous nationalism. Each country likes to see its own gold rush as the one that either shaped those that followed or epitomized all the rest. In *The Gold Crusades: A Social History of Gold Rushes, 1849–1929*, Douglas Fetherling takes a different approach.

Fetherling argues that the gold rushes in the United States, Canada, Australia, New Zealand, and South Africa shared the same causes and results, the same characters and characteristics. He posits that they were in fact a single discontinuous event, an expression of the British imperial experience and nineteenth-century liberalism. He does so with dash and style and with a sharp eye for the telling anecdote, the out-of-the-way document, and the bold connection between seemingly unrelated disciplines.

DOUGLAS FETHERLING, D.Litt., is a poet, fiction writer, critic, and small-press publisher. He is the author or editor of fifty books, mostly in the fields of literature and culture. He has been the literary editor of two newspapers, the *Toronto Star* and the *Kingston Whig-Standard*, and writer-in-residence at Queen's University. He divides his time between Ontario and British Columbia and writes a column on books and ideas for the *Ottawa Citizen*.

DOUGLAS FETHERLING

The Gold Crusades:
A Social History of
Gold Rushes, 1849–1929

REVISED EDITION

UNIVERSITY OF TORONTO PRESS
Toronto Buffalo London

© University of Toronto Press Incorporated 1997
Toronto Buffalo London
Printed in Canada

ISBN 0-8020-8046-4

Printed on acid-free paper

Canadian Cataloguing in Publication Data

Fetherling, Douglas, 1949–
    The gold crusades: a social history of gold rushes,
    1849–1929

    Rev. ed.
    Includes index.
    ISBN 0-8020-8046-4

    1. Gold mines and mining – History.   2. Gold mines and
    mining – Social aspects.   I. Title.

    TN420.F47 1997   338.2′741′09   C97-931092-X

This book was first published in 1988 by Macmillan of Canada.

University of Toronto Press acknowledges the financial assistance to its
publishing program of the Canada Council for the Arts and the Ontario
Arts Council.

For Roy MacLaren

Contents

THE GOLD CRUSADES

Introduction:
Gold Crusaders

Gold is not found in quartz alone; its richest lodes are in the eyes and ears of the public.   *The Notebooks of Samuel Butler*

When Fred C. Dobbs first meets Howard, the old prospector, in B. Traven's novel *The Treasure of the Sierra Madre*, it is in a Mexican doss-house where Howard is reminiscing about his experiences searching for gold: 'See, I've dug in Alaska and made a bit; I've been in the crowd in British Columbia and made there at least my fair wages. I was down in Australia, where I made my fare back home, with a few hundred left over to cure me of the stomach trouble I caught down there. I've dug in Montana and in Colorado and I don't know where else.'[1] In his film version of the novel, John Huston added Honduras to the recitation. He might just as easily, and more accurately, have mentioned South Africa or New Zealand or northern Ontario, though the gist was sufficiently clear. Such lists are evocative of certain truths, the first of which is that there was once such an ailment as gold fever and that it was pandemic throughout Western society. One of the reasons it had such considerable force is that it was compressed into a relatively short span of time. There was a period of only eighty years, from 1849 to 1929, when much of the industrial world went slightly mad at the very idea of a gold rush and people of nearly every nationality created a mass movement by the simple exercise of their ambition.

The first great gold rush, at all events the first truly international one, the first one on a grand scale and with the gaudy characteristics by which later ones could be identified, followed an accidental discovery of gold in a California stream-bed in 1848. The Australian rush of the early 1850s drained away much of the force of the California one and quite a few of its participants, a portion of whom then moved on to New Zealand. Of those who chose to remain in North America, many became caught up in the rush to British Columbia later that decade or else retraced their steps eastward through the goldfields of Nevada and Colorado. In each case, individuals were forced to push on not only because the excitement died down but also because big business moved in once the easiest pickings were exhausted. Prospectors were thus stalking-horses for financiers, to whom they stood in opposition but with whom they nonetheless had a certain quality in common – call it adventurous self-determination.

One of the early rushes, the one in Nevada, offered a twist on the usual formula because it was silver that was discovered first. Silver provided the impetus, and also the money, to exploit the subsequent discoveries of gold. So, too, in South Africa in the 1880s. The diamond rush there made a few people wondrously wealthy rather than fulfilling the usual promise of making a multitude moderately so. This led naturally, it might even be said inevitably, to a South African gold rush, capitalized and exploited by those same few. Those who had taken part in previous rushes, as well as those who participated in later ones, were known by humble, self-deprecating names – forty-niners, diggers, fossickers, dry-blowers, sourdoughs. In South Africa, however, the principal players were known by a lustier term, randlords.

By the 1880s, the whole business of gold rushes was tied to the promotion of mining companies, and this was even more so in the early 1890s when there was another gold discovery in Australia, this time across the deserts in the western half of the continent. Looked at in the context of these events, one of the reasons the Klondike stampede of 1898 was so dramatic and far-reaching – why it excited the popular imagination in a way no other had done since California fifty years earlier – was that it was not dominated by corporations. It could still offer people an achievement in proportion to their luck and the amount of hard work they could do with their own hands. This came as a refreshing message indeed in a world where modern industrialism was a vice whose jaws were already beginning to clamp shut on people's lives. The first gold rush in 1849 was to some extent made possible by the Conestoga wagon. The final one, in Canada during the 1920s, was barely practicable because of and in spite of the aeroplane. When the gold rushes began, passports were seldom necessary. By the time they ended, all the relevant countries had long since introduced the personal income tax.

The desire for gold, of course, has often been a motivating force in history. Its presence was the engine that drove the European explorers, as in the Americas. In the seventeenth century, among the English settlers in the New World, 'there was no talk, no hope, no worke, but dig gold, wash gold, refine gold, load gold.'[2] Yet even in Brazil, there was no actual gold rush in the modern sense. Geological and social conditions were ripe for one, but the gold was a monopoly of the Portuguese Crown and the Roman Catholic Church. Nor was there a gold rush in Russia in the middle of the nineteenth century, as there might otherwise have been, because all facets of production were controlled by the tsar. A true gold rush is predicated on freedom of movement coupled with a belief in people's ability to better themselves. That is to say, gold rushes were an outgrowth of British liberalism, related to both free trade and home rule. They were another product of the industrial revolution and another reaction against it: the expression of some impossibly arcadian ideal. The fact that the gold rushes were in essence a British imperial phenomenon shines through, time and again, as one examines the events in sequence. But because the first rush took place on American soil, America set some of the terms for the later ones. Gold rushes came to mean overreaching, greed, fraud, violence – in any case, no cooperative commonwealth. California also demonstrated the abuses that the subsequent rushes sought in various ways to correct.

In most if not all cases, gold was known to exist in the area long before the fact came to seem so important. Gold alone was not enough to ignite a stampede, since stampedes required so many other ingredients, the conjunction of certain sociological, technological, and economic factors. The economy had to be expansive, for instance; people who might rush after gold to escape recession found that gold rushes suffered just as any other sector of the economy did. There also needed to be a frontier on which the gold rush might take place so as to underscore the sense of exploration and adventure as well as to provide the necessary remoteness. This dovetails with the fact that gold rushes happened at specific technological times. They became possible only when news of the discovery could be spread quickly (and exaggerated), and they were no longer feasible once it became too easy for more than a small minority to attend. Thus, the pell-mell convergence on California in 1849 occurred well into the age of the telegraph and steamship but before the transcontinental railway. It was easy for news to get out into the world but difficult for the world to act on it, so the pressure mounted. Most important of all, each such event demanded a new generation of people prepared for overexertion in their efforts to express themselves through their actions and make their part in the mass hysteria a type of individual statement.

The notion of going off to seek an overnight fortune became a rite of passage for successive generations of young men – and young women as well, though their numbers and importance were downplayed until recently. That the Klondike gold rush began to peter out immediately after the Americans declared war on Spain is appropriately symbolic, not merely coincidental. The type of people who would put enthusiasm ahead of good sense to climb the Chilkoot Pass were also those who would put glory above morality to charge up San Juan Hill.

The fact that the gold rushes, whether on American or British soil, were made up of people from so many different countries strengthens rather than weakens the argument that they were an outgrowth of Manchester liberalism. So does the fact that the same people went to rush after rush, whether they had been successful or not, regardless of what stratum they occupied in the gold rush hierarchy. The seven sons of a Swiss trader named Meyer Guggenheim operated in mineral rushes up and down the Pacific Coast of the New World from Alaska to South America. At quite the opposite economic pole, so did Major William Downie, the prospector for whom the gold rush town of Downieville, California, is named.[3] But most left no records. There is little way to trace the uncounted and anonymous people who followed the same route or some less linear course, zigzagging from continent to continent and hemisphere to hemisphere. There were thousands of Germans and Britons who went gold hunting in America and thousands of Americans who went to Australia. Part of the same phenomenon were the South Africans and Canadians who went to Australasia and the Australians who went to South Africa, and so on. Every nationality was represented in some measure, except at first the Japanese and Russians, who were forbidden to emigrate, but towards the end they too were in evidence.

J.A. Froude, the English historian, wrote of this floating population of international prospectors as 'Bohemians of all nations,' but that is not quite what they were. A few were charlatans and rogues and many were exiles. A good number were highly educated. But most were exceptional only in the degree of their ordinariness. They were not necessarily shiftless. What linked them was the rootlessness born of optimism or else the disillusionment which they perhaps blamed on conditions at home or the imperfections in the age. They convinced themselves that what ailed them or their time could be cured by ascribing a higher purpose to their common journey.

In all of this, they were very much like the Crusaders. It was the journey and the process, not the destination or the fact, that lent mystery
and substance to the Crusades, which were in effect a single discontinuous event spread over a long period of time and affecting politics, economics, diplomacy, and even fashion, just as the gold rushes did. There
never was a Holy Grail; like the mother lode, it was a romantic conceit
of the Victorians, in whose increasingly secular age Jerusalem had to
share the map with Eldorado, a place equally unattainable. It is this element of fancy and at times self-delusion that essentially distinguishes
the true gold rushes from the lesser and more recent events to which
they sometimes bear a superficial resemblance. In chasing a vision, the
gold crusaders were hurrying away from civilization, though they often
found that elements of it had preceded them to their destination; unlike
the contemporary army of young people who work the world's oil rigs,
they were not seeking a job for wages.

The subtext of the gold rushes – and a pretty obvious one it is – was
the struggle between order and chaos, between authority and lawlessness, between British and American ideas of society. California was the
American paragon of free enterprise at its freest. The miners quickly
formed committees, believing this the God-fearing and democratic thing
to do, and then used them to enforce a reign of terror on whoever dis-
agreed with them or on whomever they disliked for racial or other reasons. California took the heritage of the New England town meeting
and created the tradition of vigilante justice. It was only the absence of
any other kind of government that prevented the California miners from
rebelling against it.

In Australia, as in other locations later, the authorities were in place
before the rush struck. They alternately suppressed news of the gold,
hoping to avert social upheaval, and encouraged its discovery with
rewards and bounties, attempting to channel the energy. The miners at
the diggings in Victoria were the usual rich mix of nationalities but with
a generous portion of the sort of Englishmen who were legatees of the
Reform Act of 1832 and no doubt had read their Cobbett religiously. A
few European radicals and some ideologically simplistic American
troublemakers added seasoning to the group. Alas, the official charged
with maintaining order in the goldfields was one of those smug, incom-
petent baronets who seemed to surface at the wrong moments in colonial history. The reforms that resulted from the inevitable armed clash
were adopted with only minor variations in subsequent gold rushes,
from New Zealand in the 1850s to Western Australia in the 1890s. The

diamond fields and goldfields of South Africa might have witnessed revolts as well but for the white South Africans' talent for repression. South Africa remained British but was dominated by Boer rather than British culture; it was certainly run for the benefit of the few mine owners, who effectively usurped the power of the Crown and used it for their own purposes. The Klondike gold rush gave a clearer message, the one crystallized by Pierre Berton: how order and good government prevailed in Dawson, on the Yukon side of the Canada–U.S. boundary, while violence and corruption were the hallmarks of Skagway, on the Alaskan side.

However much a country may acknowledge the international character of the horde, the nations in which important gold rushes took place tend to view their own as an isolated event that arose naturally from indigenous conditions. This may be because each stampede had such a dramatic effect on the development of its host country. When the gold rush crusade is considered in such terms, it is usually regarded as an American convention, because the United States claimed the first one and so many Americans were involved in all the others, and because gold rushes are bound up with the idea of the frontier and its unbridled individualism. But this need not be the case. For example, viewing Canada's or Australia's gold rush as part of a cumulative worldwide development is not inconsistent with the best instincts of Canadian or Australian nationalism. On the contrary, to do so is to seek in the British imperial heritage a rival form of individual expression that is separate from Americanism and untainted by it.

One result of the contest between authority and libertarianism is that there seem to be no wholly accurate figures on how much gold was produced, let alone how much it was worth, either at the time or in today's purchasing power. In California, no reliable figures were kept because there was no need to keep any. In Australia, the Crown imposed an export duty on the newly mined gold, all but guaranteeing that small producers would evade the duty as much as possible. In the Klondike, the Canadian government tried to exact a royalty on the gold at source; one observer has suggested that only one troy ounce in ten was ever declared. In any event, each country's system of tabulation seems to have been different from that of the next, making cumulative totals even more suspect. There have been conscientious attempts to make sense of this confusing situation. In the case of the earlier gold rushes, the most notable effort was that of the mining engineer and economist Alexander Del Mar (1836–1926), though it helps put his work in perspective to remember that he was one of the thinkers whose views on monetary policy influenced Ezra Pound.[4] In any case, the volume of gold was staggering. In the 1960s it was estimated that at least 20 per cent of all the gold mined since 1500 has been wrung from the earth during only fifty years' worth of gold rushes in the nineteenth century, and this was very probably a conservative estimate.[5]

It is also important to bear in mind how significant was the number of people involved in the rushes. Again, there are no totally reliable figures, not least because there was no way of judging the amount of overlap from one rush to another. But the fact, for instance, that about 35,000 of the people who set out for the Klondike in 1898 and 1899 actually arrived there represents, for that day, a very large peacetime migration indeed. The equivalent figure in today's terms would be at least ten times as many or even more.

Just as they all sprang from the same motives, so too did the gold rushes have certain stock characters in common. There was the individual who made the initial discovery or claimed to have done so. He was often a local farmer or labourer who found gold while looking for something else or by tripping over it or accidentally breaking open a rock to reveal a yellow seam. Similarly beloved in the popular culture were the few prospectors who did in fact make fast fortunes and live like kings, at least for a time. Then there were the entrepreneurs who supplied necessities or essential services to the miners. There are stories galore of dry goods dealers who built dynasties and of widows who grew rich baking pies with dried apples to sell to the scurvied prospectors at outrageous prices. Another type of people were the holdouts from previous frontiers who rushed backwards through time seeking to relive their exciting youth. Yet another were the literary observers who originated so many of our preconceptions about the subject. They range from Bret Harte, who made a career of recreating the atmosphere of gold rush California, to – of all people – Frank Harris, the author of *My Life and Loves*, who used his all too facile pen to publicize the silver mines of Cobalt, Ontario.[6] The one who had the most varied and intimate acquaintance with the topic was Mark Twain. Fittingly, for someone so enthralled and repulsed by his own gilded age, he was a prospector in the 1860s in both California and Nevada, seeking gold and silver by turns, though his efforts were unsuccessful. Later, as a highly paid world traveller, Twain followed the old forty-niners' trail across the Panamanian isthmus. He visited the goldfields of Australia as well as those of South Africa. 'I had been a gold miner myself,' he wrote of a trip to Johannesburg, 'and knew substantially everything these people knew about it, except how to make money.'[7]

But with each successive rush, the atmosphere became more desperate, and these and other characters seemed to grow more pro forma and clichéd. It is significant that although the California gold rush was rich in genuine folk song, the others produced only doggerel, both written and sung, of the sort associated with Robert W. Service. This is not to deny that the gold rushes in general still have an important place in a wider tradition, where they serve as a kind of historical crosscurrent. As the historian and critic George Woodcock observed, 'It is not without reason that such historical phenomena as the gold rushes have found their place in folk literature, for their essential elements of tragedy, disillusionment and plain human folly and weakness inevitably appeal to ordinary men as a representation in reality and on a grand scale of their own everyday fantasy lives.'[8]

The Gold Crusades: A Social History of Gold Rushes, 1849–1929 was first published by Macmillan of Canada in 1988. For this new edition, I have corrected errors, taken advantage of later research in the field, and updated and expanded the 'Essay on Sources.'

1

The California Delusion

Because of the chaos and disorder that the world's first great gold rush unleashed, it is tempting to think of California in 1848 as a still paradisical place untouched by lawyers or venereal disease. To be sure, it was an outlying part of that Mexican empire which United States citizens had been rubbing up against for generations in their appetite for ever more land. There were so many Americans in the province of Texas by the 1830s that they had broken away and established an independent republic, biding their time until Texas could be admitted to the union. When admission came to pass in 1845, Mexico and the United States went to war. Humiliated by an American invasion, Mexico capitulated and ceded to the United States all of what now constitutes California, New Mexico, Arizona, Nevada, and Utah. But the Treaty of Guadalupe Hidalgo was still nine days from ratification when James W. Marshall (1810–85) made his famous discovery in a northern California creek in three feet of water.

Marshall was typical of the footloose Yankees who were beginning to find their way to California, the kind who always lurk on the ragged edges of civilization. A product of the same sociological gene pool that gives America so many of its musical prodigies and lone assassins, he was born in New Jersey and trained as a carpenter like his father and was forever moving a little farther west. In search of greater prosperity, he went first to Indiana, then to Illinois, then across the Missouri River where the states ended and the great undifferentiated territories began. In 1846 he was part of the northwestward migration to Oregon after the British had been forced to give up their claim to the territory. Later he drifted southward and into the employ of John Augustus Sutter, another archetype.

Sutter (1803–80), a short, elegantly dressed man who was nonetheless bigger than life, claimed to be a retired captain of Louis Napoleon's esteemed Guards. In fact, he was a defaulting Swiss businessman eager to stay one continent ahead of his creditors. He was also what a later age would recognize as a textbook psychotic: charming, cunning, ruthless, romantic, and given to grandiose schemes in which other people, blinded by the reflection of his enthusiasm, could find themselves believing.

In 1839 Sutter had persuaded the Mexican authorities to grant him an enormous tract of land in the valley of the Sacramento River, and he had recently been currying favour with the American military occupiers; for his grant had become New Helvetica – an enterprise distinguished from the utopian communities so common in the 1840s not by its lack of idealism but by its abundance of profit, all of which devolved on Sutter. He had first built a large, thick fort (similar to the reconstruction that is a popular tourist spot today). This fort became the hub of a vast self-sufficient spread, part ranch, part farm, part trading post, with generous woodlots and vineyards. Sutter employed about two hundred persons in all, mechanics, artificers, labourers, and the like, and was forever making improvements and additions. On the agenda for 1848 were a gristmill and a sawmill, the latter to be built forty-five miles up the American River in the Coloma Valley. James Marshall was in charge of the party of eighteen men and one woman contracted to do the work. On 24 January 1848 he was standing in the stream where the fifteen-foot tailrace was to be located when he saw an object glittering in the late afternoon sun. It was half the size of a pea and looked to him like gold.

As Marshall told the story through later years, his heart began to pound with excitement. But he was sceptical of his good fortune. He tried to convince himself that it might be nothing but iron pyrite, the 'fool's gold' on which the plot of so many Hollywood westerns would turn one day. No miner himself, he nonetheless knew that pyrite was brittle, whereas gold was malleable. This was malleable. 'Boys,' he told the others, 'by God I believe I have found a gold mine.'

Sutter was no geologist either, but when Marshall brought him the evidence of his discovery, he knew how to test the sample with nitric acid and measure its specific gravity by weighing it in water. For gold, after all, was the one universal medium of exchange, worth sixteen dollars a troy ounce by common consent regardless of the issuing authority (hence the sentence in *Moby-Dick*, published in 1851, when Captain Ahab, in nailing a coin to the mast of the *Pequod* as an inducement to catching the white whale, touts it as 'a Spanish ounce, worth $16'). In 1848 Spanish and other coins still circulated freely in the United States as legal tender alongside the products of the official mints. Like the Ahab of fiction, Sutter was a man of commerce accustomed to dealing with many different currencies and thus had some knowledge of metallurgy. Nor was he a stranger to politics. He urged Marshall to keep the secret to himself as much as possible while he, Sutter, negotiated a treaty with the Yalesumni Indians that would give him title to all the mineral wealth in the area. But when the American military governor at Monterey, Colonel R.B. Mason, learned of the proposed treaty, he pointed out that only the federal government could enter into such agreements. By that time, word had slipped out.

In a drinking place in the village of San Francisco, Sutter's messenger fell into conversation with a man named Isaac Humphreys, who had taken part in the closest thing to a gold rush that had so far taken place in the United States. In the 1830s gold had been discovered in what seemed substantial amounts in an area shared by South Carolina and Georgia, and this had since been the country's chief gold-producing region. Individual initiative and risk had certainly been important factors, but the event had for the most part been a regional phenomenon, though in 1838 the federal government had established mints at both Charlotte, North Carolina, and Dahlonega, Georgia, to regulate the supply and convert it to its most useful form. Humphreys thus had actual experience of gold mining and would be credited with introducing to California the technique of panning for gold that he had first witnessed in Georgia. Humphreys, Marshall, and Sutter resolved to go into partnership to exploit this new discovery, though in the end the discovery exploited them. Humphreys has disappeared from the historical record, but Marshall was soon charging collectors for his autograph and vainly trying to convince people that he could locate gold by occult means. Sutter saw his empire trampled and became a favourite figure of tragedy to artists as diverse as Blaise Cendrars, the French poet, and Sergei Eisenstein, the Russian filmmaker.

The fact that there was gold in California was hardly unknown. A British geologist had written of it as early as 1816, and ten years later, in a well-known incident, a trapper had discovered tangible evidence in the Sierra Nevada. Certainly, the Native population had been trading gold to the whites for almost as long as the whites had been there to receive it. But California was no longer a somnolent appendage of Mexico, largely unpopulated and with no economic reason for existing; it had suddenly become part of the land of hope and bunkum and would soon become a state if the thorny question of whether it should be admitted as a slave state or as a so-called free-soil one could be settled. Once the secret had a wide enough audience, gold became an excuse as well as a reason for going to California, a reward for having completed a symbolic journey as well as the fulfilment of what American expansionists called their country's manifest destiny – to fill up all the space available to them between the oceans.

Only seven years before Marshall's discovery, the journalist Charles Mackay had published *Extraordinary Popular Delusions and the Madness of Crowds*, a book to which generations of economics students and stock-market forecasters have looked for fundamental wisdom about why people move en masse towards delusion, as during the South Sea Bubble or the Tulip Panic. Most of Mackay's examples were historical, and he himself was a Scot, but he had toured extensively in the United States and was writing at a time when the type of mass hysteria for which he was seeking an explanation seemed all too likely to break out again without warning. It was the age of P.T. Barnum and the penny press, when the popular culture was particularly debased and people were given to consensual credulity. Physically, too, it was a period of mass movement, much of it caused by war, revolution, famine – and land hunger. Culturally and politically, then, the timing of the California discovery was propitious.

Despite the attempts at secrecy by Sutter and his confederates, the news spread, but it was at first acted on rather slowly and then only along easy geographical lines. It travelled with sufficient speed to heighten the sheer excitement of the discovery – but slowly enough so that maximum pressure was permitted to build. Just as an explosion is a fire that burns its fuel too quickly, so a gold rush, it became apparent, was a wave of immigration with cataclysmic force. Certainly, it was no mere economic happening but something extraordinary that changed the country's demographics and the way its citizens, and those of many other nations, would think.

The autobiography Sutter later wrote is notable for its elaborate falsehoods, but he did record accurately (because he remembered it with excruciating clarity) the day the news he already knew became common knowledge at his fort: 19 May 1848. The result was a frenzy bordering on mêlée, as the usual crowd that had business there – traders and workmen; deserters and ne'er-do-wells; Californios, both adopted and native – scrambled for the spot on the American River where the gold had been found. Men whose tools were their livelihood dropped them in a trice and departed. At the nearest permanent community of any size, San Francisco, the scene was similar. The plank-and-canvas village on its beautiful harbour had a population of perhaps eight hundred and fifty, of whom six hundred were adult males: Sandwich Islanders (as Hawaiians were known), Mexicans, Yankees, and three Chinese. Two-thirds of the inhabitants immediately pulled up stakes and set out for the scene of the discovery. In the short term the hysteria would make Sutter rich supplying the miners, and in the long term San Francisco would boom and blossom. Building lots that a few months earlier had changed hands slowly at $22.50 were being grabbed up at $3,000.00 in September.

In its early months the stampede succeeded in roughly mapping out the boundaries of the gold-bearing area. The gold was there all right, in nuggets, flakes, and dust, deposited here and there beneath rocks and in the concave sides of sand and gravel bars, anywhere that its own weight had combined with some obstacle to deposit it, as through the centuries, the millennia, wind and water had freed it from the mountains of hard rock. There was immediate talk, as there would be in later rushes, of finding what the Spanish called *la veta madre* and the Anglos called the mother lode: the large reserve, still embedded high up in the mountains, from which erosion and gravity had wrenched the scattered pockets being found down below. But in fact the gold being found here was itself the mother lode as it had been redistributed into a billion components down through geological time.

Initial investigation centred not only on the American River but on others such as the Yuba and the Feather, which likewise drain into the Sacramento. This was high, wet country, from two thousand to five thousand feet in elevation, with angular rock formations topped in thick vegetation. The streams were gravelly, and the gravel beds changed frequently in the heavy winter rains. But by following their luck, the early risers of the California gold rush quickly moved south into the antithetical country along the Calaveras, Mariposa, and Consumnes rivers, which feed the San Joaquin. Here the hills are lower, rounder, and rolling, the vegetation more brown than green and closely cropped. The area became known as the Dry Diggings, because there was no rainy season to drive the prospectors away, nothing to prevent them from sifting through a promising gravel deposit until they had extracted every fleck of gold that it was within the scope of their technology to find. By July, there were two thousand whites digging for gold alongside an equal number of Native Americans; it is believed that they took in an average of fifty dollars a day each – at the time a windfall sum indeed. By August, when there were already six thousand miners in place, the exploitation of the southerly territory was under way. From top to bottom, the two goldfields stretched about one hundred and twenty-five miles. Later, as hordes from around the world began cramming in, it was to extend two hundred and fifty miles.

Compared with the events of 1849 and 1850, the gold rush in its first year, 1848, was a small regional affair. Yet it began to assume its later characteristics almost from the beginning. The venue was not simply the new American nation but the frontier of the new American nation, and not the more familiar though constantly revised interior frontier but the farthest edge of a society whose civilized centre visitors found rough and grasping enough. The first murder recorded at the sawmill where it all began took place as early as October 1848. Over the life of the rush there were hundreds of murders and lynchings by both individuals and gangs. In his diary, William Perkins, a Canadian who kept a store at Sonora in the Dry Diggings, confided his horror at every act of violence by the uncouth Yankees; but by spring 1852 even he had grown callous, complaining that there were 'no fights, no murders, no rapes, no robberies to amuse us' as the California rush wound down.

One obvious reason for the violence was California's ambivalent political position. Mexican colonial law was no longer applicable, but it had not yet been replaced by anything more than the incentive to good behaviour implicit in the presence of the American military garrison at Monterey (most of whose members, many hundreds in all, deserted as soon as possible to join the gold seekers). California in fact had no criminal code. Later, when the society turned its attention to such abstract matters, it was at first only to establish a slate of mining laws.

A second reason for the violence was the multicultural composition of the cast, which marked the gold rush almost from its first moments. There was already tension between the native Californios and the interloping Yankees (the latter being a term which those in the southern states applied inclusively to all in the northern states, who in turn restricted it to people from New England). There was tension between the whites, particularly the Yankees, and the aboriginal population. It became good sport as well as shrewd business to cheat the Native Americans of their hard-dug gold, obtaining handfuls of it in exchange for a handkerchief, six ounces for a string of glass beads, and, in one instance, nine ounces for a blanket. Even these figures were often fraudulently measured 'digger ounces' (the whites often called the Natives diggers because of their custom of digging for roots and other food). As the news of gold was loosed on the world, more and more nationalities and races were added to the mélange.

When the first surge of chaos swept Sutter's Fort in May 1848, a courier left for the U.S. capital with dispatches and a month-old San Francisco newspaper telling of Marshall's discovery. He rode far south to Santa Fe, then east and north to the Missouri River, whence a series of steamboats and lesser conveyances eventually deposited him at Washington City three months later. In the interim, news travelled by the routes of least resistance. The Sandwich Islands, with which California had substantial trade, had the word in June, and by 19 October ships were leaving Honolulu for the Golden Gate filled with Kanakas, as the Polynesians were often called. Oregon, itself a potent mix of accents and pigmentations, soon began to export its residents as well. In September, around the same time that the first specimens of gold from California reached the eastern United States, a full shipment of newly mined gold was unloaded in South America, at Valparaiso, touching off an immediate exodus of Chileans. They and those who followed them played such a large part in the California rush that Yankee miners used the term *chilenos* for all Spanish-speaking miners except Mexicans, who were themselves arriving in substantial numbers. Most of the Mexicans were from the mining state of Sonora. Their practical experience gave them an advantage and also earned them additional suspicion and dislike.

News stories describing the discoveries in California, reports that were usually inaccurate, often garbled, and frequently not believed, began appearing in the larger population centres in the eastern part of the North American continent and in Europe. The institution of the exchange desk allowed a newspaper proprietor to copy without charge articles printed (perhaps already at second hand) by some faraway editor. Consequently, one can follow the dissemination of the story, much as epidemiologists can chart the progress of a disease from town to town, noting the way it skips some communities and infects others. New Orleans, St Louis, Cincinnati, Pittsburgh, Philadelphia, Baltimore, Boston – the general direction was both south to north and west to east, followed by early outbreaks in Britain, where one newspaper satirized the American expectation of becoming 'rich as Jews,' and in France, where California became a national obsession. James Gordon Bennett's New York *Herald* confidently pinpointed 24 November 1848 as the day 'California gold fever broke out in New York.' That was ten months

after Marshall's chance discovery and just as the heavy rains were ending the mining season in the northern fields. A miniature polyglot
republic had already sprung up in California. San Francisco had not
only boomed but had already burned down for the first time, as it
would often do in the future. Fortunes had already been made, lost, and
supplanted by others.

At this stage, it is estimated, $3.7 million in gold had been extracted.
There is no sensible way to translate such a figure into contemporary
terms; to come up with a meaningful comparison, one must rely on specific examples of the purchasing power of the American dollar in 1848.
An unskilled labourer might earn one dollar a day, a semi-skilled
labourer such as a cook, two dollars. Common soldiers were paid seven
dollars a month, and members of Congress eight dollars a day. But all
this soon changed, at least in California. Inflation was such that there
were soon $200 whores. And such was the social climate that their pres-
ence was taken as another sign of abundant civilization.

The California gold rush was in various ways and to varying degrees
both the natural extension of the mainstream society and a reaction to it,
but what was that society like? What were people running from and
towards, and what did they pass through along the way?

Andrew Jackson, who as an adolescent had taken part in the American Revolution and lived long enough to be photographed by Mathew
Brady, came to the end of his long life in 1845. His America, however,
was still in session, considerably more industrialized, and a good deal
less hopeful than it had been during his presidency, but it was instantly
recognizable by its democratic clatterwhacking (a characteristic 1840s
word) and its joyful discovery of the frontiers of its own ignorance. Yet
there was something hollow in the boasting about equality, for although
it had never been easier to become rich, there had never been a time
when the classes of people were farther apart materially. It is easy to
overlook this fact now, because our picture of that nondescript decade
comes mainly from the few cultural artifacts that have survived as representative, which is not to deny that they can be instructive.

The America of the forty-niners, as those massing in response to the
spreading news were called, was growing rapidly, and getting worse
perhaps as it grew larger. Between 1840 and 1850, the population rose
from 17 million to 23 million, with as many as 750,000 immigrants in a
single year. Ireland and Germany were the largest depositors. It was,
accordingly, the time of the Know-Nothing Party and various other
nativist and generally anti-Catholic movements. The veterans of the

rebellion against Britain had all died off, so now the past could be wor-
shipped free of contradiction. George Washington had long since
become the most common subject of icon makers. Noah Webster, who
died at the end of the decade, had reformed spelling because he considered the spelling of many American words tainted with evil Britishness.
Politicians made hay, as they might have said, in the belief that divine
providence had intended the United States to occupy the entire North
American continent. This patriotism was a current running through the
culture from the highest plane to the most popular. 'Eastward I will go
only by force,' said Henry David Thoreau, 'but westward I go free.' It is
difficult to understand now why he, along with Ralph Waldo Emerson
and Margaret Fuller, seemed so radical when their ideas were conducive to the usual American patriotism. No doubt it was partly their sim-
plicity that many found attractive, for the 1840s was an ornate time
above all else.

The old indigenous folk culture seemed to be ending, replaced by the
America of Stephen Foster and of Currier and Ives prints, with their
absolute nostalgia. In architecture, the Greek Revival was petering out
and the Gothic style was asserting itself. It was the bumptious America
that Charles Dickens reported on in 1842 – the America that read
Dickens, even his sour and unflattering *American Notes*, viewing his
novels as the pinnacle of a form of fiction that was immensely popular,
even the three-decker romances written by lesser practitioners. The
books were no longer made from rag paper but were printed on paper
made from acidic wood pulp, as everywhere craft was dismissed in
favour of mass production, which inevitably meant mass consumption
as well. The first woman's suffrage convention was held in 1848, though
it was largely a curiosity. Women were the recipients of society's most
elaborate discourse, and in any case abolition and prohibition were the
biggest and noblest causes. People liked what came to be called genre
painting, often of idealized landscapes or historical themes. In the ten-
year period just ending, the number of pianos in the country had doubled – there was now one for every 2,777 Americans – but the fashion-
able instrument was a variety of small reed organ called the melodeon.
Objects were becoming fancier on the surface but cheaper in substance,
as were ideas and the quality of public life.[1]

A New Yorker named William Miller had come to believe that the
Second Coming was scheduled for 23 April 1843 and had convinced
untold thousands of others. The non-appearance of the Messiah caused
Miller to recast his figures and arrive at a date in October 1844, but again

his followers emerged unsaved, and many Millerites moved to Canada. Other people were not so easily disillusioned. It was the age when science and chicanery were never very far apart, the age of hydropathy and phrenology, the time when Abraham Lincoln and Karl Marx were beginning their careers in an atmosphere in which people evinced a strange desire to believe and to move forward, at least geographically if in no other way. Any such series of dominoes is more difficult to trace than the recurrence of the gold rush story in first this newspaper and then the other, but it is important to grapple with it. To be able to conjure up some of the atmosphere in which news of the gold rush was received makes it easier to accept the extraordinary hardships people underwent in order to be part of the event.

In January 1849 the New York *Herald*, which had been running California stories for the past five months, caught what must indeed have been the flavour of life for many thousands of people:

All classes of our citizens appear to be under the influence of this extraordinary mania ... Will it be the beginning of a new empire in the West; a revolution in the commercial highways of the world; a depopulation of the old States for the new republic on the shores of the Pacific: the future alone can answer ...

In every Atlantic seaport, vessels are being fitted up, societies are being formed. All are rushing towards that wonderful California which sets the public mind almost on the highway to insanity. Look at the advertising columns of any journal and you will find abundant evidence of the singular prevalence of this strange movement. Every day men of property and means are advertising their possessions for sale in order to furnish themselves with means to reach that golden land ... and every day similar clubs of the young, educated and best classes of our population are leaving our shores. Poets, philosophers, lawyers, brokers, bankers, merchants, farmers, clergymen – all are feeling the impulse.

The societies or clubs referred to were the most common means by which Americans participated in the rush. They were bands of from half a dozen to as many as one hundred and fifty men, generally in their twenties and usually friends, colleagues, or schoolmates, who pooled the cost of the journey with the promise of dividing the profits as well. A few were standard commercial ventures with charters, shares, and rule books, but most were the simplest sort of congregation dignified by a loftier name. Some bore vaguely patriotic titles such as the Kit Carson Association or the Peoria Pioneers, while others depended on bravado: the Bug Smashers, the Helltown Greasers, the Rough and Ready Company of Boston. Most strove for a dignified commercial resonance combined with some indication of the place of origin: the Connecticut Mining and Trading Company, the Mutual Trading Company of Salem, the Nantucket and California Company. It seemed that nearly every sizable town had at least one. Certainly, all thirty-one states were represented, and some Southern prospectors brought their slaves with them, though it is thought that most of the estimated one thousand African Americans who joined the rush were, in the contemporary phrase, freemen of colour. The communal spirit as an impetus for going to California did not in practice usually mean cooperative effort once a group arrived there. The exercise of American individualism was an essential part of the attraction, and what native ideology failed to achieve was accomplished by greed and fear.

What we see on looking back at these events is not only the expression of everything that was American but also the establishment of the worldwide gold rush phenomenon, which was later played out on various scales in many parts of the world and was subtly transmuted into other types of rushes – silver, diamonds, real estate, even rubber and other unromantic commodities. Wherever a gold rush happened to be taking place, that was the amorphous frontier suddenly given shape. All the stock characters normally attracted to the outer edge of society by a kind of centrifugal force were to be found in abundance. California almost instantly became a festival of three-card monte dealers, thimble-riggers, and larcenous persons of all sorts, along no doubt with bigamists, ship jumpers, embezzlers, and other fugitives.

With them came the true entrepreneurs. It was a truism in all later rushes, predicated on the California example, that the people who made the real money, or at least kept it, were those who provided the miners with goods and services. This fact obtained even at the lowest level, with women who washed shirts for more than the shirts had cost in the east. One favourite device, which persisted until the age of radio, was the importation of big-city newspapers, hired out in reading rooms or simply read aloud to an audience at so much a head.

In later rushes some famous fortunes – the type that continued to ennoble the families of those who had founded them – were indeed made by miners who had discovered fabulously rich deposits. But in California, some of the best-known fortunes were made in businesses related to the mining, not in mining itself. Mark Hopkins, for instance, joined the New York rush to California in January 1849. By 1853, he was running a Sacramento ironmongery with Collis Huntington, who had wheeled and dealed his way to California, making money by supplying fellow passengers as he travelled. Before the decade was out, the two had joined Charles Crocker and Leland Stanford in financing a railway, the Central Pacific. The four names were virtually synonymous with West Coast capitalism and were known collectively as the Big Four. Heinrich Schliemann (1822–90), a German who had already made one fortune as a Crimean War profiteer, made a much larger one in California by buying and selling gold dust after the rush ended. He used his wealth to become the father of modern archaeology, the discoverer of Mycenae and (so he believed) Troy.

Gold seekers of other nationalities were no less captivated by the promise of wealth, and many had the additional motivation of peace and political freedom. In 1848 there had been revolutions in Paris and Vienna and serious rioting in Berlin, Prague, and Madrid, all linked by the gathering force of working-class discontent. The Italian city states were particularly volatile; there was a revolt in Rome (the Pope had fled), Lombardy had been besieged, and there were insurrections in Palermo and Venice. Although the situation had settled, it showed no signs of resolution or relief; nor did the prognosis for Ireland, where the great potato famine was killing hundreds of thousands. In such cases it is difficult to ascertain how many immigrants were primarily gold seekers and how many of the gold seekers had originally come for other reasons. Before 1848 was over, goldfield maps had found their way to the Prussian town of Exin and were seen by the three Levinsky brothers, Louis, John, and Mark, who rose to prosperity in California by selling work clothes, boots, and bowie knives to the miners. Louis Levinsky was the grandfather of Alice B. Toklas, who was born in San Francisco in 1877 during the long afterglow of the gold rush legend.

For many, the gold rush could not have chosen a better time to take place, politically or economically. The Taiping Rebellion in 1850 sent many Chinese, whom the whites nearly always called Celestials, to the goldfields. It is recorded that two thousand departed in one forty-eight-hour period. In 1849 alone, forty-five ships carried them from Canton or Hong Kong, where speculators advanced many of them the fare in return for equity in their success – which a number did in fact achieve, usually through service work. Most Chinese immigrants to California, one hundred thousand of them by 1882, were from Kwangtung province in the south, which was overpopulated even with the famine that had devastated it throughout the 1840s.

British involvement in the rush was substantial, but it has generally been impervious to study because the English and the Scots, and no doubt the Welsh and Irish too, tended to participate as individuals without banners or corporate groupings. Indeed, many of the British-born gold seekers had already been in the New World for years. By contrast, the French reacted to the news with clearly identifiable unanimity. The whole notion of the gold rush appealed to the French character. At one point, three French plays set in the California diggings were being performed in Paris theatres simultaneously. One of them was written by the young Jules Verne. Like British and Latin American financiers, the French made a considerable effort in underwriting the rush, with ventures such as La Compagnie californienne. By 1850, about two thousand Frenchmen were in the mining region, including some adventurous members of the aristocracy, though most were ordinary citizens, often winners of a lottery created for the purpose of sending people to California. Just as some individuals used the rush as an escape from politics, the French government used it as a convenient method of disposing of potential political troublemakers. France also sent San Francisco its first contingent of European prostitutes, who followed the arrival of others from various South American ports.

In February 1850 two hundred Australian women (many of them what the rhetorical style of the day termed soiled doves or fallen angels) disembarked, complementing the Australian miners, who had been a forceful presence in the rush since shortly after the departure of the first ship from Sydney on 9 January 1849. One of the most fearsome sections of gold rush San Francisco was Sydney Town, the home of an organized gang of former Australian convicts called the Sydney Ducks, who terrorized citizens at will. Then there was Little Chile, which was not so little. There was even an Arab quarter. Although there was much racial tension, culminating in the anti-Chinese rioting of subsequent years, many have testified to the comparative absence of religious intolerance. The respect achieved by Levi Strauss (1829?–1902), a Bavarian Jew who arrived early in 1854 and dealt in dry goods, and who eventually won immortality as the creator of strong canvas trousers for miners – Levi's – attests to the lack of anti-Semitic feeling. As the years wore on and recollection lengthened, along with pride at having been a forty-niner or almost one, there was perhaps a clearer sense that these diverse characters had been engaged in some great enterprise together. As they looked back on their youth, the urge to read brotherhood into the fact that they had even survived the harrowing journey must have been irresistible.

The common estimate is that 80,000 people joined the California gold rush during the calendar year 1849. This figure becomes impressive in the context of what seems to modern eyes a rather thinly populated world. London, for example, had a population of about 2,600,000. New York City had about 500,000 residents, Montreal only 175,000, and Sydney 60,000. Perhaps one-third of the gold seekers (often called argonauts in the neoclassical style of the nineteenth century) came from outside the modern United States, though there are no reliable or even plausible statistics to support such a conclusion, which is merely an impression derived from reading many different texts.

The route they travelled depended on where they started from, of course. Although conditions were often appalling, the voyage from Canton was straightforward if arduous. South Americans and Australians likewise followed clearly understood sea lanes. Britons and other Europeans first crossed the Atlantic and then started the rush to California on an equal footing with the Americans themselves, which is one reason why it is difficult to disentangle the various nationalities. These argonauts, who made up the majority, had basically three choices. First, there was the overland route from the eastern half of the continent, across the prairies, the Rocky Mountains, and the desert; second, a long voyage from some East Coast port, round the tip of South America and up the Pacific side; third, the shorter sea route, which involved an overland journey across Central America to the Pacific and thence by ship to San Francisco. Each had several variants. Each also had both advantages and serious drawbacks. Even more than what the miners did or failed to do in California, the means by which they arrived there is what really defined them, in their own eyes and in society's.

People who lived in an eastern seaboard city or within a short distance of one tended automatically to think of the gold rush as a seagoing venture. But the demographic centre of gravity had been moving westward for years, and the entire American Midwest, the area drained by the Great Lakes, was comparatively well populated if still largely agricultural. People who had already made one or more westward hops in their lives or who had grown up rootless in families that progressed westward by staccato leaps over two or three generations were those most inclined to act on the glittering suggestion of California. Thus, the overland routes were the choice of the greatest number of argonauts.

A popular tune was so much on the lips of the argonauts that it became a virtual theme song of the gold rush. This was Stephen Foster's 'O Susannah,' in which the speaker implores his intended not to cry for him though he has 'gone to California with a banjo on my knee.' The gold rush produced a rich legacy of folk songs, some of which help recreate the atmosphere very vividly.[2] The United States as such ended in a ragged western border, which for much of its length followed the Missouri River. Thus, people in the western territories beyond could sing 'What Was Your Name in the States?' Its message was that asking a person what name he or she had lived under in the United States was at best impolite and probably dangerous. There are circles in which such restraint is still proper etiquette, to be admired in others and carefully repaid in kind.

The song 'Sweet Betsy from Pike' speaks even more eloquently. Over-landers had to make their way westward by a debilitating combination of coach, steamer, and embryonic railway in order to arrive at the staging areas on the Missouri. They found that the people already established there, being the leading edge of the westward expansion, were the most susceptible to pushing on still farther; ergo, Sweet Betsy from Pike. Pike County, Missouri, on the Mississippi between St Louis and Hannibal, Mark Twain's birthplace, was one of the jurisdictions suddenly almost depopulated by the gold rush fever. The ballad tells of one Betsy who set out for California with a lover named Ike and with two yoke of oxen, a spotted pig, 'a tall Shanghai rooster,' and a yellow dog, only to run low on rations, encounter cholera, fight off aboriginal warriors and the covetous looks of polygamous Mormons, and nearly die of thirst in the alkali desert. 'Such songs,' noted the folklorist Alan Lomax, 'were performed by the professional entertainers who toured the gold camps, and were circulated in the little pocket song books of that day.' An even more gruesome retelling of horror is 'The Days of '49.' In its more familiar versions, it is obvious that it was written looking back at the crossing many years later, though the time elapsed can have done little to make the experience more benign.

During the previous few years a near-constant stream of caravans had taken American settlers across the West to the newly acquired Oregon Territory. It was a journey fraught with danger. Memories were still fresh of the infamous Donner Party, an emigrant train that had been trapped in the deep snows of the Sierra Nevada in 1846, whereupon at least one of the survivors had resorted to cannibalism. Now what was called the Oregon Trail was suddenly given new utility as a route to California. The trail began at Independence, Missouri, and ran northwest through the present-day states of Kansas, Nebraska, and Wyoming, where the two great halting places and provisioning stations were Fort

Laramie and Bridger's Fort. Both were heavily fortified trading posts, similar to Sutter's.

For much of the first half of the journey, the Oregon Trail ran roughly parallel to the Mormon Trail, which had been used by Brigham Young and his followers as a route to the Great Salt Lake, where Providence had commanded them to settle. But once the Rockies hove into view, the two trails diverged. The Oregon-bound settlers pushed sharply north, through present-day Idaho and Oregon until they reached the Pacific. But the gold seekers took advantage of a newly found network of river valleys to move southwest out of the Rockies, across the desert of what is now Nevada, and into the mountains once more. They entered California (a promised land even richer mythologically than that of the Mormons) at the crook of the modern state's eastern border, between those two gamblers' shrines, Reno and Lake Tahoe. Beyond that, the mountains conveniently drained into the Sacramento River. A variant route cut more sharply south, near that other future monument to avarice, Las Vegas, and terminated in southern California.

Individual parties often made improvised trails for short distances or were led along them by unscrupulous managers. But what became a complicated system of trails and subtrails used by wagon trains overlapped a great deal and shared a general pattern – west from Missouri, then north, then west again for a thousand miles or more, then south over the mountains. The wagon trains are one of the most durable images of the American West as it has come down through the filter of the cinema and other entertainments. The silhouette of the long caravans is true enough to history, though like all silhouettes it depends on shapes without detail. In fact, the wagons were more sturdily made than popular culture usually suggests. They were actually goods wagons, with wide rims and strong axles, stout beds, and slightly upturned ends to facilitate fording streams and rivers. There are places on the American prairie where one can still see the ruts carved by these heavy slow-moving vehicles. Also, the wagons were often brightly painted. Their billowing canvas canopies were likely to bear names such as Wild Yankee (indicating that the owner was a New Englander), Hoosier (a resident of Indiana), or Buckeye (someone from Ohio). Some variant of 'California or Bust' or 'Ho! for California' was seen frequently. One more literary wagon carried the wording 'Pilgrim's Progress – California Edition.'

The wagon trains might contain just a few wagons or as many as a couple of hundred, as small groups of argonauts from one region, commonly with a democratically chosen captain, merged with others, usually under the stern discipline of a professional guide who demanded so much per wagon. Photographs of the forty-niners show how the men often began their quest in natty dress commensurate with their high hopes, only to discard waistcoats and bright shirts for simpler more utilitarian garb in the goldfields, just as soldiers' uniforms tend to become less elaborate as a war wears on. The same reductive process was apparent with regard to the argonauts' goods from home. The trails were littered with fancy mining equipment, costly furniture, pianos, and other heirlooms. They were also dotted with the hastily dug graves of those who died along the journey. One traveller of 1850 counted 963 graves and guessed that more than 4,000 others had died on the journey. But no one knew for certain.

The most horrible danger was cholera, though there were other diseases no better understood. Even scurvy loomed large in the primitive conditions found on the trails. Some people perished of thirst in the deserts, just as others froze to death in the mountain snows. A few were killed by Native people, and wagon groups often formed into an oval when they camped for the night. But Hollywood has generally exaggerated the danger of such attack on the Oregon Trail and related routes. This particular danger was greater by far on the more southerly ones.

Those taking the southern routes across the continent most often began their journey at Little Rock or Fort Smith in Arkansas, cutting west across what is now Oklahoma and the northern panhandle of Texas to emerge at Santa Fe in New Mexico, a well-established centre of trade and exploration. From Santa Fe, they dropped south, almost to the Mexican border, then threaded through the mountains via the Guadalupe Pass to Tucson, and thence farther west to San Diego or Los Angeles, both small sleepy places. There they could take ship to San Francisco if they did not wish to proceed farther by land. The entire trip required exquisite timing so that one's animals could enjoy suitable grazing throughout the journey.

Here, as with the Oregon Trail, there were many alternate routes which together comprised a fabric of opportunity. One confidence man promoted a scheme to lead a party across Texas from the Gulf Coast to El Paso. He turned out to be a notorious scoundrel, one Parker French, who purchased supplies with forged letters of credit and walked a precarious path between his clamorous creditors and the Comanches, whose hostility was a forceful argument against such an itinerary. Three of the men died in the desert; others had to break away for their own safety, relatively speaking, and escape into Mexico. One who did so was a Canadian argonaut, Charles Cardinell. He was subsequently wounded and two of his companions were killed when assaulted by French, who turned renegade and began attacking the people he had already duped.

A still more southerly route was taken by J.W. Audubon who, like his father John James Audubon, was a naturalist and painter. He descended into Mexico from the tip of the Rio Grande, moving laterally across Monterey, over the Sierra Madre Occidental, and up through Sonora. Bandits as well as Native people preyed on the pilgrims in this rough country. As well, the Mexicans as a rule were unfriendly towards the American victors of the late war who had occupied their country. Of course, the Americans generally did little to alleviate others' dislike of them, mocking the local customs both secular and religious. One aspect of the young Audubon's hardship was that he lost the many paintings and sketches he had made along the way.

The safest avenue to the goldfields was also the longest – by ship around the Horn. The first ships full of argonauts from New York, 360 of them, left as early as the first week of October 1848, and almost at once the trade became a sizable one at all the eastern ports. At Baltimore, Philadelphia, and Boston, and at whaling towns such as Newport and New Bedford, ships were refitted for the passenger trade and others were hurriedly built. In the last month of 1848 and the first two months of 1849, 11,000 people sailed from such cities in 178 vessels, and the numbers grew at a crazy rate.

By sea, San Francisco was 17,000 miles from New York. The average fare was $200, though some travellers paid as much as $1,000. The New York newspapers were filled with column after column of miners' goods being offered for sale. In fact, it seemed that anything, however unsuitable to the climate or to mining, could be sold to unsuspecting argonauts if the word California was included in its name. Equally prominent were the long lists of ships available to carry people to the land of wealth. Many were chartered for the California trade by middlemen similar to today's travel agents or by the larger argonaut associations. On these, the passengers could have at least some voice in decisions. Not so on the independent ships, many of which lured travellers with false claims about the promised food, accommodation, and even seaworthiness. There were a number of shipwrecks on the way to the goldfields, though perhaps few considering the circumstances.

The historian Hubert H. Bancroft, who in subsequent years was honoured for his West American Historical Series in thirty-nine volumes, was a twenty-year-old fortune seeker in 1852 when he saw survivors of the steamer *North America* reduced to living on local charity after the ship had run aground at Acapulco. Mass meetings were called in San Francisco and in Stockton to the east in order to raise money for their assistance. The *North America* was one of half a dozen Pacific steamers operated by Commodore Cornelius Vanderbilt, the New York capitalist, and in time four of the six were lost to such accidents. Later, as the rush ended, the steamer *Central America*, of another line, perished off South Carolina on its return from California full of newly wealthy miners and their gold. It was not located by salvagers until 1987.

The ships that were lined up at their berths hurriedly taking on passengers for the West were sailing vessels and steamers of both types, side-wheelers and the newer screw propellers. Many of the square-riggers were slow and clumsy, but choosing steam over sail was not a guarantee of greater speed; the steamers were no match for the clipper ships. One of these marvels of marine engineering sailed from Baltimore to San Francisco in 112 days instead of the usual 150 for ships generally. One of the most famous, the *Flying Cloud*, completed a New York to San Francisco run in only 89 days. In one 24-hour period, it logged 374 nautical miles, more than any other sailing ship or any steamer had ever done. Yet choosing even a clipper ship did not assure one of better conditions, for conditions were universally appalling.

There are a great many accounts left by passengers who had to endure brackish water, rotten meat, and wormy meal or biscuit. One of the staple dishes was lobscouse, the fish stew familiar to generations of British sailors. Other dishes were apparently American creations, probably with African antecedents, such as dandyfunk. This was a sweet cake made with animal fat and molasses. One recipe book intended for argonauts admitted that it was not 'a high-toned dish.' Just as unappealing as the food was the harsh discipline or, worse, the lack of concern bordering on deafness of the masters of some ships. Two vessels out of Boston, the *Capitol* of 700 tons, which first made for California in January 1849, and the *Duxbury*, an old three-masted schooner that got under way the following month, both experienced passenger uprisings on their maiden voyages over the quality of the food, and both ships saw several more revolts before the gold rush ran its course.

Some ships were restricted to males, while others were not. At this remove, it is difficult to say which choice offered the greater potential for discord. There was a certain amount of larceny, to judge from the diaries and journals, but there were so many complaints about boredom that the discovery of a thief may have been a pleasant diversion. Music was common on California-bound ships, and several were equipped with printing presses, like the ocean liners of a later day, and issued newsletters. In one case, a book was written and printed during the voyage round the Horn.

The first leg of the journey was the longest and least hazardous, ending with a stop at Rio de Janeiro. Rio was a place of such size and sophistication that even the influx of as many as a thousand Americans a day did not result in the social disruption seen in some of the smaller Latin ports. Passengers were not allowed to disembark the first day their ship was in the famous harbour. When they did, they were met by a swarm of touts and shills, but they were generally fond of the place if also, in many cases, repelled by evidence of the slave trade. A certain percentage landed in difficulty by challenging the local culture and duly found themselves reposing at the American consul's if they were lucky – or, if not, in jail or even the penitentiary. Some vessels made short stops at Bermuda or other places before calling at Rio, and others made freshwater calls at Santa Catarina Island after Rio. But as a rule, Rio was the only preparation for the ordeal of the Horn itself where, for days or more commonly weeks, contrary winds could keep ships from proceeding at all while the sea threatened to drive them on the rocks. The Straits of Magellan presented a short cut in terms of miles but at the price of increased danger.

On the northward journey up the west coast of South America, the ships had a variety of ports of call, including the Juan Fernandez Islands, which had seen few visitors since Alexander Selkirk, the original of Robinson Crusoe, had been marooned there one hundred and fifty years earlier. However, the primary point of contact, analogous to Rio, was Valparaiso. This was a community of 30,000 people, where the Americans swaggered through the streets with what a local newspaper complained was 'the aplomb of ancient Romans in a foreign city,' and where they were obscenely overcharged for all goods and services, illicit and otherwise. Despite their behaviour and the reaction to it, they seem in general to have liked Valparaiso inordinately, these thousands of Americans who arrived every day in steamers and schooners and barges and brigantines. Once Valparaiso was cleared, the sea traffic became heavier, for the Cape Horn stream merged with that from the Central American route.

As mentioned above, the Cape Horn route to San Francisco from New

York was 17,000 miles, required perhaps 150 days to complete, and cost $200. By comparison, the route from some southern U.S. port via Panama was only 5,000 miles and required a mere 35 days, but it cost $400. So Panama it was for those who could afford to pay twice as much to get there in one-fifth of the time. This was the choice of the most devout optimists, who believed they could make a fortune if they got to the diggings early or were in place to supply those coming later. It was also the choice of businessmen, professionals, aristocrats of every description, and of course gamblers and adventurers, who found New Orleans a convenient point of departure.

Here most of the hardship was not to be found on the ocean. The voyage across the Caribbean to Chagres in Panama was often uncomfortable but just as often it was uneventful. Chagres was a low, malarial place where the argonauts bartered for passage up the Rio Chagres, often in large punts called bungos. Thus, fleets of steamers and sailing vessels, most of them operated on a high-volume commercial basis and many, as the rush progressed, purpose-built, were disgorging hundreds of anxious passengers at a mere coastal village that had few facilities and rather primitive prospects for connecting transport. Here began the enormous bottlenecks associated with the Central American routes. Most of the passengers going up-country, a journey of sixty miles and some days' duration, had never seen such lush tropical country or heard such bizarre jungle noises. If the adjective California was used by U.S. merchants to make goods more attractive, the adjective Panama seemed to take form by itself to describe an all too necessary article, the Panama hat, as well as a dreaded scourge, the Panama fever.

The end of the river journey was at one or the other of two interior villages, Gergona or Cruces, where the travellers transferred to mule or some other conveyance for the fifty-mile trip through the rain forest to Panama City on the Pacific side. A railway to the capital, though under consideration since 1846, was not completed until 1855, by which time the California gold rush had lost its momentum. Even during the gold rush period – a full two generations before the Panama Canal and the creation of the U.S. Canal Zone – the country took on the appearance of an American colony. Argonauts of other nationalities remarked on the strange American custom of plastering the jungle trails with signs and advertising messages. Panama City, where the jam of people was most obvious and tempers were shortest, became a kind of American commercial and political centre. The steamers which people expected to meet them did not arrive or were already full. There arose a complex system of bribes, and the wages of sin was crippling inflation. People sold their jewellery in order to survive or took to robbery. A satirist in a New Orleans newspaper contended that there were six hundred American lawyers in Panama City. Twenty-one of them, he claimed, were hoping to represent California in the U.S. Senate and fully four hundred in the House of Representatives. Seventeen had already begun campaigning for the governorship. The other Central American route was through Nicaragua – from San Juan del Norte on the Caribbean, up the San Juan River about one hundred and twenty miles, and an equal distance across Lake Nicaragua. From that point, only a range of mountains a dozen miles deep stood between the traveller and the Pacific.

It was not long before argonauts in general, regardless of their financial standing, began to choose the Central American routes over any of the overland ones, for although Panama and Nicaragua were as different from one another as any of the other choices, each had its own advantages. The Nicaraguan route was marginally the less arduous of the two, if also longer by approximately one hundred miles. It had originally been developed by the Spanish as a means of foiling the British privateers and pirates who preyed on their Panamanian trade, a fact that is of ironic interest in view of the diplomatic consequences inflicted on Central America by the gold rush.

Perhaps the most hated historical figure in Central America is William Walker (1824–60), an American forty-niner whose name, despite a 1987 Hollywood film, has been almost totally repressed in the political consciousness of the United States. Walker was a sometime lawyer and sometime newspaper editor from Tennessee who became the country's most notorious filibuster, a term that had not yet come to mean the use of long speeches to delay passage of legislation but instead corresponded to what would now be called a soldier of fortune. In 1850 Walker turned up in San Francisco, as so many thousands did, leaving no record of his journey, no connective tissue joining the eastern portion of his life with the western. Unsuccessful in gold, he tried to make his name as a military man. In 1853 he invaded Baja California and declared himself its president. Later he became even bolder and tried to annex Sonora, but Mexican troops forced him back across the border, where the American authorities acquitted him of violating the neutrality laws. The next year, 1855, he invaded Nicaragua and made himself president, being recognized as such by the U.S. government. But he was to run afoul of one American in particular, Commodore Vanderbilt, a powerful man indeed.

Long before the gold rush began, Nicaragua was a familiar topic of conversation in Europe and America because of diplomatic clashes between the United States and Britain, and sometimes France too, over the possibility of a canal through Nicaraguan territory as a way of expediting the lucrative China trade. Britain had built up its military and political presence in the region with an eye to the future, but it was Vanderbilt who emerged the probable winner. He had made his fortune, one of the most famous of the nineteenth century, through shipping monopolies in New York, but rivals had stolen a march on him in the gold rush passenger market. Once his own fleet was in place, however, he resolved to build a canal and corral the market himself. He acquired a charter from the Nicaraguans containing a clause that would give him the same exclusive right to build a railway or road if a canal proved impractical – which turned out to be the case, particularly after he failed to interest British investors in a joint venture. What he ended up with was a line of inland steamers operating on the San Juan River and Lake Nicaragua at immense profit. When Walker came to power and began interfering with the vessels, his fate was sealed.

Vanderbilt and his agents orchestrated an overall Central American uprising against Walker, the relics and memory of which are today almost sacred. But Walker escaped and repeatedly tried to finance new invasions of Nicaragua until 1860, when he was captured by the British, who turned him over to the Hondurans, who promptly had him shot. For his part, Vanderbilt sold his Nicaraguan enterprise in 1858, when the California gold rush was becoming history, and concentrated on U.S. railway deals. The route atrophied over a long period until Panama became the accepted place to cross Central America on interocean travels of whatever kind. When Mark Twain crossed from the Pacific to the Atlantic in 1866, he used the Nicaragua route, but he found it marked by a sense of desertion by the forces of history. Nevertheless, his description of the country and its inhabitants jibes well with the accounts of the forty-niners. San Juan del Sur, the Pacific port ballyhooed by Vanderbilt, was only 'a few tumble down shanties – they call them hotels – nestled among green verdure.' In his observation that the Nicaraguan women were 'buff colored, like an envelope' and 'singularly full in the bust, the young ones,' one finds the same lack of sensibility with which the forty-niners so often failed to ingratiate themselves in Latin America.[3]

What did the argonauts find after crossing the American wastelands, sailing round the Horn, or traversing Central America? What of San Francisco, the disembarkation point for most of the forty-niners and the permanent home of many? And what of the mining camps themselves? A wealth of eyewitness testimony exists, much of it as delightfully implausible as it is demonstrably true.

There is something marvellous in the way that cities, as ongoing institutions quite apart from the particular people who at any given time inhabit them, retain the essential characteristics of their founders. One theory of urban history in the United States holds that all but three of its large metropolitan centres were settled by Puritans, the descendants of Puritans, or pseudocavaliers of the Old South who were the cousins of the descendants of Puritans, with their asceticism and schematicism, their indifference to history, and their hatred of culture. The three exceptions are New Orleans, Chicago, and San Francisco, cities settled by adventurers, grafters, bummers, gin sots, flimflam artists, prostitutes, and the like, and so not altogether coincidentally the most livable cities in the country. Of the three, San Francisco retains an extra dimension, perhaps in part because it is a perishable city whose 1906 earthquake and fire underscored the transient character with which it was imbued by the gold rush, with its thick air of accidental cosmopolitanism in a frantic hurry.

The literary historian Van Wyck Brooks succeeded better than most in creating a thumbnail sketch of gold rush San Francisco:

One saw vaqueros swinging their lariats and capturing wild bulls in vacant fields where houses rose overnight with families in them, and a visitor in 1851 counted six hundred ships in the bay, almost as many as one saw at the port of London. It was generally supposed in San Francisco that traders and farmers of the plodding sort had stopped on their westward march to people the prairie, while only the adventurous and reckless pressed on to the coast, and the cosmopolitan crowd brimmed over with soldiers of fortune from every land, prepared for any turn of the wheel of fate ... There were ex-doctors sweeping the streets, ex-ministers who were gamblers, bankers and Sicilian bandits who were waiters in cafés, lawyers washing the decks of ships and penniless counts and marquises who were lightermen or fishermen or porters. The spirit of chance prevailed in the town, where the topsy-turvy was almost the rule, though the golden stream from the miners flowed all over. The streets swarmed with spendthrift miners who came down the rivers from their gulches and bars, hungry for any amusement and ready for a fling ...

The French, who were sometimes prodigal sons, opened hotels and kept cafés, where the songs of the boulevards mingled with the ballads of the miners, and they were croupiers in gambling-houses that had an air of Paris with their plush and their diamonds and velvet and chandeliers and mirrors. Saloons and parlours, both public and private, blossomed with Second Empire furniture, which the pioneers imported from France in the flush of their wealth, a note of gilded opulence that flourished in the Chinese restaurants too and the shops that recalled the bazaars of Canton and Peking. There were theatres and fandango-houses, pantomimes and minstrels, for the town was full of actors, musicians and dancers ... Two daily papers were printed in French and others in half a dozen tongues, and some of the San Francisco editors were adventurers from the South who had much in common with the gamblers and the errant young Frenchmen. For many French royalists and Southerners alike were stirred by the dreams of an earlier age.[4]

Most, however, were stirred by the bald desire for quick riches, a goal that could be approached from many angles in an instant society where everything was in short supply, including intangibles such as accurate information.

When news of the gold strike broke, the only book on California in print was *Two Years before the Mast* by Richard Henry Dana, Jr, published in 1840; but emigrant guides and accounts by individual travellers speedily began to appear as soon as the publishers could oblige the eager public. The first was *The Gold Regions of California* by G.G. Foster, which was rushed out in the last few days of 1848. The first substantial work and surely one of the most reliable was Bayard Taylor's *Eldorado, or Adventures in the Path of Empire*, published in 1850. Taylor (1825–78) was already a popular travel writer, the author of *Views Afoot; or Europe Seen with Knapsack and Staff*, and a general literary man who, significantly, had published poems about California, an imagined California (and whose translation of Goethe's *Faust* is still used today). Eager for dependable intelligence, Horace Greeley dispatched him to the afflicted area of society to report for the New York *Tribune*. This proved to be a fortunate choice, for Taylor was a dependable reporter and a clear and colourful writer.

He left New York in June 1849 with a kit that included a barometer, a thermometer, a revolver, and a sturdy suit of corduroy. At Chagres, his first landfall, he found no hotels or restaurants but was quite content, delighted in fact, to eat a humble meal while squatting on an earthen floor in the company of dogs and swine. Taylor loved roughing it, though he did not let his enjoyment distort the information he gathered with such voraciousness. One of the virtues of his book, which was popular in Britain and was almost at once translated into German, was the way he ventured far afield to describe, as much as possible, the customs and culture as well as the topography and climate. He arrived at San Francisco to find a city that 'seemed to have accomplished in a day the growth of half a century.' He continued: 'On every side stood buildings of all kinds, begun or half-finished, and the greater part of them mere canvas sheds, open in front, and covered with all kinds of signs, in all languages.'

Many of his fellow passengers 'began speculation at the moment of landing,' and Taylor himself was caught up in the game. He had used copies of the *Tribune* to cushion his belongings in transit, and soon he found that he could sell these old newspapers to a wholesale newsagent at a mark-up of 4,000 per cent. Clearly, the local economy was out of control. Taylor noted that seventy-five prefabricated sheet-iron houses had been shipped from China, along with Chinese carpenters to assemble and finish them, because this was cheaper than trying to purchase materials or engage labour locally. The classic example concerned the miners' long underwear. In the gold rush economy, such manufactured goods were absurdly expensive to replace, yet it cost more to have them laundered in California than to have them done in either the Sandwich Islands or China; so a ship engaged for that purpose would arrive with its hold full of fresh linen for all. 'There is no end,' Taylor later observed, 'to the springs of labor and traffic, which that vast emigration to California had set in motion, not only on the Pacific coast, but throughout all Polynesia and Australia.'

In the main square he saw a fifteen-by-twenty-five-foot tent hotel called the Eldorado which, he reported, was rented to local gamblers for $40,000 a year. Like many other observers, he was especially impressed by the rents – as much as $160,000 a year, according to most accounts, brought in by gamblers at the leading hostelry, the Parker House (named after the Boston hotel where the form of bread known to Americans as Parker House rolls originated). Being a landlord was one of the surest ways to wealth. A man had died the previous autumn owning a little property though technically he was a bankrupt with debts totalling $41,000; before his will could be probated, the heirs were raking in $40,000 a month in rents. Goods had to be auctioned on the wharves as soon as they arrived because the cost of warehousing inventory grew by 20 per cent a month. The ships that brought them lay idle, because entire crews deserted once they reached San Francisco and there was no one to replace them. Servants earned $200 a month. On another side of the square from the Parker House stood the United States Hotel; Taylor saw children outside it with knives earning five dollars a day digging flakes of gold.

The rapid growth distorted time. Taylor left San Francisco to travel about the diggings for four and a half months, and when he returned the tent city had become a metropolis of fifteen thousand people. 'All the open spaces were built up, the canvas houses replaced by ample three-story buildings, an exchange with lofty skylights fronted the water, and for the space of half a mile the throng of men of all classes, characters, and nations, with carts and animals, equalled Wall Street before three o'clock' when the stock market closed for the day. By the time Taylor returned to the East, overland across Mexico to Vera Cruz, and set foot again in New York in March 1850 after an absence of eight months, the population of San Francisco had doubled once more, to thirty thousand.

But it was the conditions at the diggings that most interested his readers, and this information is at the heart of Taylor's book. His own experiences centred on the northern diggings near Stockton, which was then a canvas colony of a thousand residents where one paid a hotelier four dollars a night to bed oneself and seven dollars a night to bed one's mule, with no food for either. The miners worked in *arroyos*, or gulches (neither term was then understood in the East), but of course only when the season permitted. Arriving at the Mokelumne Diggings on the river of the same name, Taylor joked that 'for the first time in several days, we slept in a bed' – the bed of the river, which at other times might be under thirty feet of water, like that of the neighbouring Calaveras.

In the dry diggings, the gold was deep. Taylor told the story of an argonaut who gave up after two days of hard digging only to be followed by a German who, after only one further day's excavation, was taking out $800 in gold daily. But here it was a question of washing gold from between boulders and breaking up layers of slate. The miners he saw used 'a rude cradle for the top layer of dirt,' followed by shovels and finally wooden bowls as the piles of dirt grew smaller and the elusive gold finer. At the first camp at which he lingered, the men were averaging seven ounces of gold a day each, so they ate very well indeed, with meals of roast turkey, fresh oysters, 'Goshen butter,' and fresh corn, beans, and peas. But the work was extraordinarily strenuous, even with the aid of a little engineering knowledge such as that possessed by a group of twenty men downstream who had spent a month building a dam: 'When I first saw men carrying heavy stones in the sun, standing near waist-deep in water, and grubbing with their hands in the gravel and clay, there seemed to me a little virtue in resisting the temptation to gold-digging; but when the shining particles were poured out lavishly from a tin basin, I confess there was a sudden itching in my fingers to seize the heaviest crowbar and the biggest shovel.'

At that moment there were, in Taylor's estimate, 100,000 men in the diggings, which had assumed many of the characteristics of small towns. Places with such names as Angel's Camp, Rough and Ready, Poker Flat, and, ominously, Hangtown were simple in their social structure. The sites that Taylor visited were being worked by Kanakas, Mexicans, and Frenchmen, as well as by Americans, some of whom hired Native people to do the heavy labour for them. Whenever the necessary combination of rumour and geology caused such a community to take form, the American inhabitants exerted their democratic tradition and elected an *alcalde*, or mayor. In practical terms, his task was to preside over the maintenance of law and order – which is to say the near-absence of both. Persons accused of theft or other such crimes were whipped or maimed, or hanged from one of the live oaks. Taylor saw one man whose ears had been cut off. Other offences arose from crooked gambling. Taylor saw one camp, a mere accumulation of tents, that boasted ten montebanks (the term, also spelled mountebank, then referred to the monte dealer's set-up rather than to the dealer himself or, by extension, any charlatan).

A number of diarists agreed with Taylor that the Mexicans were the heaviest gamblers, though he met many non-Mexicans who had the bug, including a man named Buckshot who made and then frittered away $30,000 and spent his evenings alone in his tent drinking champagne. The Mexicans at this time occupied the role later suffered by the Chinese as the main targets of racial suspicion, intolerance, and violence. Taylor stated that there were about ten thousand Mexicans working the diggings and that they sometimes formed bands and 'took possession of the best points' and so were ordered to leave. A figure from California folklore, though he was real, was Joaquin Murieta, a Robin Hood who avenged his brothers by robbing the Anglos. One of Taylor's discoveries was a group of Australians, 'a company of Englishmen, from New South Wales,' who had no experience of mining and had been in California for only a fortnight. One of them, who had spent seven years in a British cavalry regiment, struck it rich and exclaimed, with that attitude always calculated to please the patriotic Americans, 'By me soul, but this is a great country!'

Hundreds of other reports and journals flesh out Taylor's picture,

contradicting it in places but confirming it in others and generally building up the pastiche until it becomes something close to reality in the
reader's mind. A dozen accounts could be cited for their importance.
That of Daniel Woods, for example, is notable because its author visited
both the northern and the southern placer districts and also because he
seemed to be genuinely repulsed by the spectacle of unbridled greed.

Woods travelled by a less common route, across Mexico. He was
armed with a letter, in Latin, from the Roman Catholic bishop of Philadelphia, from where he reached San Francisco in 145 days. After a brief
stay there, which occasioned in him the customary comments ('Gambling seems to be universal. Rents are held at the most exorbitant
prices'), he proceeded to the American River, 'induced to come ... by the
accounts we received of the success of two brothers – Jordan – who, in a
few weeks, made $3000 here, and are now on their way home.' Woods
met men from Cincinnati, Ohio, as easily as he met a man from Tahiti
and 'a Dutchman' (by which he probably meant a native German). He
acquired a quick grasp of mining and its technology, but his own
rewards were never large and reached a nadir of $1.25 a day. He polled
the loose associations of miners, or companies, that came his way and
compiled detailed statistics. Of thirteen such groups, four had made
nothing and the others had taken in from $1.09 a man per diem (the low)
to $20.37 (the high). Woods found that the average amount taken in by
fifty-six miners was $3.26 daily.

He grew rich in experiences, however. At one point he actually witnessed the birth of a town: 'Upon a bar above our dam some miners
lately met with some success. Rumours of this success, but much exaggerated, were circulated. Ounces were reported pounds. The change at
once was magical. Trading tents, the signs of rival physicians, eating
and gambling booths have sprung up, and the noise and confusion of a
large village are heard.'

Travelling by mule train to the Tuolumne River diggings in the southern district, Woods was washed out and spent a night broke and alone
in driving rain under threat of an imminent bear attack, until rescued by
a kindly Canadian named George Islip who took him in for a week.
Woods seemed all the more eager to record the good samaritan's name
because he had met with little other kindness or pleasantness in the gold
country. He was horrified to hear that the California agent of the Rothschild banking family was rumoured to be sending $17,000 a month in
winnings to sanctuary in Europe on behalf of some local gamblers. His
common sense balked at paying ninety-six dollars for a pair of boots

and two dollars each for pickles. But his greatest sadness was reserved for his fellow men. He wrote: 'Old friends meet, exchange a few words, and hasten on to the shrine of *Mammon*. Multitudes die, the waves close over them, and they are forgotten. It can hardly be supposed that people come to California *to live*, since they are here only *preparing to live* – much less do they come here to *die*. I pray that my life may be spared till I return to a land of friends, and where man is united to man by the sympathies of life!'

On his return to the East after sixteen months, Daniel Woods apparently became a priest. He was already that staple figure of every American city and town between 1850 and 1900 – the person who had gone to California as a young man, had endured all the hardships and processed them as nostalgia, and may or may not have struck it even a little rich (probably the latter), but who did have one advantage over the unbidden: he had at least taken part. Another typical argonaut was a young Canadian who left his home in March 1849 accompanied by a friend. He resolved to keep a journal in the form of letters to his family in the Niagara region. He found himself waxing homesick, though the feeling soon passed, what with the limitless fund of new sights and experiences, which he described in what he no doubt considered a racy and sophisticated style:

You cannot imagine how anxious I am to hear from you and home – we hear News now from the diggins but near as we are we cannot rely on the reports, though all agree in one point: that gold is abundant. Kelly was to write home from Sutters just to let you know of our safe arrival and I shall deluge you all with letters as soon as I arrive in town, but I must not *yell* yet for we are not quite clear of the woods today [and] I shall be out with my old Rifle as I have not yet lost all hopes of plugging a Bear or a deer before we get through.

This afternoon I saw the first gold washing – it is not unpleasant in good weather. Two men were working with a rough made box or rocker, and on washing and sifting over about 20 Barrels of sand and gravel they obtain from 12 to $15 in fine gold, the largest piece worth about a dollar or more. So you see we are exactly in the gold diggings and no mistake. No Six ways about it. No denying it. No getting round about it, but no use knocking at de door for you ain't good looking and you can't come in.[5]

It did not take long for the gold rush dream to emerge in idealized form in popular culture. Bret Harte's works were only the beginning. Cheap fiction was riddled with the theme for years to come. Stewart

Edward White, who was not born until 1873, wrung profit from hugely successful novels such as *Gold* (1913) and *The Forty Niners* (1918), but then the atmosphere lingered for a long time at many of the actual sites. Robert Louis Stevenson spent the summer of 1880 in the California gold country, an experience that produced his book *The Silverado Squatters*. Perhaps most telling is the fact that both G.A. Henty, the arch-inspirer of British imperial youth, and Horatio Alger, Jr, who prepared American boys for careers as go-getters, both traded successfully on the California gold rush, an event that might almost have been created for their purposes.

2

The Crown and the Southern Cross

After America itself, Australia was the place most affected by the gold rush of 1849. One estimate suggests that one Australian resident in fifty departed for California. Among the emigrants was Edward Hargraves (1816–91), who arrived at San Francisco in the summer of 1849, prospected unsuccessfully, and returned to Sydney in January 1851. One year later he ignited the great Australian gold rush, a remarkable event which had the effect of reversing the depopulation trend with a vengeance. In the decade to follow, New South Wales doubled in population while Victoria, the other gold rush colony, grew sevenfold.

Many of the newcomers were Californians, or at least veterans of the California gold rush, for as the crusade continued to gain momentum the roster of its personnel took on a cumulative character. By far the greatest number of Australian prospectors were British, however. In 1852, for only the second time in the history of the empire, emigrants to Australia, sixteen thousand miles from London or Liverpool, outnumbered emigrants to Canada, a mere four thousand miles away (87,881 to the former; 32,873 to the latter). Even more Chinese went to Australia than had gone to California, only to meet with the familiar persecution. The imperial connection meant that some nationalities unseen in the American rush were important to the Australian one. Within a few years came a considerable number of Muslims, mostly from Karachi, who were mistakenly called Afghans (it was they who introduced the camel to Australia for use during the Burke and Wills expedition). But the principal and all-pervasive difference between the two great gold rushes was the effect they had on the host countries.

Although it provided the model for the society of present-day California, the 1849 rush merely confirmed the existence of intrinsically American traits and attitudes. By contrast, the Australian rush actually created modern Australia. Pards, as the California miners liked to address one another, were never really partners, but in Australia mates were mates indeed: independent, self-reliant, self-respecting, egalitarian, and energetic. At mid-century, Australia was a land of convicts and sheep ranchers, a mainly rural society whose economic development effectively ceased when one penetrated more than a few miles beyond the southeast coast. By the time the rush came to an end ten years later, wool had resumed its dominance, but economically as well as demographically the colonies were a healthier mix. Sydney and especially Melbourne were important urban centres set at one end of a continent that was better understood and at least somewhat more evenly settled. And the day was drawing close when Australia would no longer be the destination for convicted British criminals. As one modern Australian historian has put it, 'With a quarter of Britain, from navvies to viscounts, clamoring for tickets to the southern goldfields, who was to think that a trip to El Dorado at government expense constituted a fearful punishment?'[1] The gold rush, quite simply, made Australia a country, though for almost half a century before Federation, it had its own distinctive brand of democracy.

Before there was gold mining in Australia there was successful copper mining, as there had been in America. In the mid-1840s there had been a copper rush in the Keweenaw Peninsula of Lake Superior.[2] Superficially it provided a small-scale preview of the California gold rush, but because of the geological nature of copper mining it was the work of capitalists and syndicates. The same was true of the Australian experience, which took place almost simultaneously and was centred on Burra Burra, north of Adelaide. The crucial difference is that in the United States, both precious and nonprecious metals belonged to those who found them, whereas in Australia and other places protected by the English common law, precious metals belonged to the Crown as treasure trove. Gold was known to exist in New South Wales, just as it was known to exist in California, long before it was, in mining parlance, proved up and certainly long before either rush began, but it existed against a backdrop of government policy on the one hand and individual illegality on the other.

In 1823 a surveyor had found gold near Bathurst, northwest of Sydney, and had duly filed his positive though undemonstrative report with the colonial administration. Like others that followed, the report was deliberately not acted upon. A decade later, a Pole, Sir Paul

Edmund de Strzelecki (1797–1873), claimed an important gold discovery, and for years afterwards he attempted to dramatize it. Although there were as yet no faraway gold rushes to excite their fears, the authorities realized that a stampede for gold would upset the status quo. Convicts made up a majority of the population. Whether they worked for families as servants and farm labourers or were under direct control of the government, their lives were not particularly pleasant, but they did have time on their hands. Both types were expected to work only until 2 PM, for example; in Australian slang, a 'government stroke' was what a person accustomed to a convict's low level of exertion could expect to suffer if he bestirred himself too vigorously. The spectacle of convicts breaking their bonds in a revolt inspired by greed was too horrible for the government to contemplate. The existence of gold in Australia was virtually a government secret, though a badly kept one owing to the nature of the gold in question.

Unlike California and even more so British Columbia, Australia was the home of free gold; nuggets, often quite large, could be found in alluvial areas just below the surface or actually lying on the ground like heavenly debris. Throughout the 1840s shepherds found these stray nuggets and sold them as contraband in the cities, particularly Melbourne. Jewellers, at least those not worried about attracting attention, sometimes displayed them in their shop windows as curiosities. Small-scale gold rushes nearly erupted several times, as in 1849 when, spurred by rumour, a group of frenzied prospectors descended on an area only twenty miles from Ballarat, where many fortunes would be made a few years later. Such pent-up energy was released only after the discovery associated with Edward Hargraves. The strict legitimacy of his triumphs is often questioned, like the legitimacy of Henry Comstock's in Nevada, but there is no denying that Hargraves was the main instigator of the first Australian gold rush and that he was rewarded handsomely.

Hargraves had come from England not as a convict or even as an assisted immigrant or settler but as a free immigrant seeking adventure and fortune. He had been or would be at various times a common sailor, a shopkeeper, a publican, and a successful beachcomber. His photographs show his mutton-chop whiskers and his even more famous girth (at the peak of his renown he weighed eighteen stone); they also suggest at least something of his character, that of a sternly practical man with a grammar-school education and a knowledge of how to act when ambition propelled him into a higher social orbit. Neither he nor the Australian friends he went with were successful in California. But while he was there he could not but notice similarities in geological conditions, and he returned home certain that he could find auriferous rock in the district around Bathurst, which he knew so well. For by now the colonial authorities had changed their tune. Having witnessed a productive lead-silver development in South Australia and in general having become more aware of the economic possibilities of mineral wealth, they had posted a substantial reward for the discovery of a payable goldfield. Hargraves was determined to claim it.

He set out on horseback over the Blue Mountains west of Sydney, travelling through Parramatta and Penrith to Bathurst, a distance of about one hundred and sixty miles. At Guyong, along the way, he recruited John Lister, the son of a local innkeeper, and with him on 12 February 1851, a month into the trip, discovered flecks of gold in Lewis Ponds Creek, a tributary of the Macquarie River. If one is to believe his memoirs, he exclaimed, 'This is a memorable day in the history of New South Wales! I shall be a baronet, you will be knighted, and my old horse will be stuffed, put into a glass case, sent to the British Museum!' But of course he knew that such minuscule evidence did not suggest, much less prove, a payable goldfield. So he took on additional help: William, James, and Henry Tom, the sons of William Tom (1791–1883), in order to cover more territory. He taught his assistants how to use a dish, what the Americans called a pan. He also taught them how to make and operate a wooden cradle like those he had seen in California. Everyone then set to work. History has never been able to state with absolute certainty who subsequently found what.

Encouraged in his ambition to capture the government's prize, Hargraves returned to Sydney in March and had a meeting with Deas Thomson, the colonial secretary, then awaited more word from up country. It is often ventured that the governor downplayed the evidence because, despite the change in policy indicated by the reward, he still feared social calamity. His official response in a report to Westminster, however, was simply to doubt Hargraves's truthfulness. 'There are circumstances attending the reports which have been made,' he wrote, 'and the appearance and character of the specimens of gold, which have been forwarded in proof of the veracity of these reports, which lead to a strong suspicion that the accounts of the nature and value of the discovery have been exaggerated.' He went so far as to suggest that Hargraves might be salting the claim with what 'is really California gold.'[3] The sensible course was to send a disinterested geologist, Samuel Stutchbury (1797–1859), to confirm the findings. Stutchbury found granular gold within three hours of his arrival at the site, and within a few days he was downright optimistic about the size of the potential goldfield. The administration saw some of their suspicions melt, which was just as well, since Hargraves was now forcing their hand.

With favourable news reaching him from the field, Hargraves ignored his confederates' pleas for secrecy and went public by means of an enthusiastic though somewhat unspecific article in the *Sydney Morning Herald*. He refrained from pinpointing the areas where gold had been found until he was assured of the reward (he was made commissioner of Crown lands and given £10,000, and twenty-five years later was granted a £250 annual pension). Once such matters had been settled, however, he provided the precise locations and left in early May for the Bathurst district. Before a fortnight had passed there were hundreds of prospectors frantically digging away. The *Morning Herald* observed, 'If we were to say that the colony has been panic-stricken, that the whole population has gone mad, we should use a bold figure of speech, but not too much to indicate the fact.' And the level of madness could only escalate very quickly in a world whose appetite for sensation and riches had been whetted by stories of California.

There were citizens who urged that the gold rush be prohibited because it would harm if not destroy the sheep business. There were even some miners who suggested that the government impose martial law so that they could all enjoy an equal chance of striking it rich under controlled conditions. Control was the important concept, to be sure. By offering the reward, the government had hoped to diversify the economy and increase revenues at no sacrifice to public order. But for all their awareness of distant California, the chaotic character of that precedent had not impressed itself on those in power in Sydney. They had to move instantly to address the gaping contradiction inherent in the fact that although all gold belonged to the Crown, the prospectors had every intention of keeping it for themselves once it was found. So it was decreed that while gold was still owned by the Crown, it could be taken legally by those licensed to do so. Effective from 1 June, licences could be procured for one pound ten shillings a month by applicants who could prove that they were not absent from jobs elsewhere. Although the fee was steep, the traffic in licences was brisk. Troopers were sent to supervise the licensing and prevent the vigilante justice so rife in California, though of course these measures did nothing to avoid the social dislocation that was inescapable in a gold rush.

Retired people and the footloose young were quick to respond to the call, but they were by no means alone. Servants fled their masters' and mistresses' employ, and independent merchants shuttered their places of business. Soldiers deserted and sailors jumped ship. The entire vigorous middle rank of society, represented by those called mechanics and artisans, was riddled with gold fever. In the westbound traffic that choked the Parramatta Road, all the strata were visible and in all manner of vehicles. A denizen of the Sydney slums pushing his few possessions in a barrow might be forced to the berm by a powerful team of bullocks – Australian oxen, usually the offspring of African bulls out of Bengal cows. Prostitutes plied their trade by night protected by the tent of darkness when not by tents of canvas. One of the most memorable sights a pilgrim could see was the land commissioner Edward Hargraves and his entourage, complete with postilions, touring the goldfields by coach.

Yet Hargraves had his swings of luck like everyone else. At first he was showered with acclaim and souvenirs. Private groups presented him with gold cups, even gold spurs. He was particularly revered in July 1852 after an Aborigine minding his sheep near Bathurst discovered sixty pounds of rich gold-bearing quartz; Hargraves quickly took charge of the madness, christening the place Ophir ('And they came to Ophir, and fetched from thence gold' – 1 Kings 9:28). But the seasonal rains turned the land of promise into one of mud and disease, and the free gold, as distinct from that one had to work to get, proved to be thinly distributed and quickly exhausted. In contrast to the California experience, few of the first wave were guaranteed an advantage by their promptness. Hargraves and his party had to flee one angry band of dejected homeward-bound prospectors who accused him of spreading false rumours. But in fact the Australian gold rush was just beginning.

As in California, one discovery had the effect of creating a climate for discovery in general. By the end of 1851, when the news from Australia began to cause frantic excitement in Great Britain, the New South Wales goldfields, supplied and supervised from Sydney, were being overshadowed by others to the south. In the part of New South Wales which had now been given separate status as the colony of Victoria, a fresh series of discoveries thrilled the public, who were soon speaking with easy familiarity of such place-names as Ballarat, Mount Alexander, and Bendigo – instant goldfield towns fed and governed from Victoria's newly chosen capital, Melbourne.

Both the Victoria diggings and those of New South Wales were close enough to their respective seaports that they had no difficulty attracting many different nationalities in short order. A Polish miner, Seweryn Korzelinski, reported:

Sometimes a happy-go-lucky German tailor, a brawny English smith, a slightly-built French cook, a Polish Jew, an American or Dutch sailor, watchmaker, [or] confectioner, a Swiss hatmaker, an impoverished Spanish hidalgo, gather near a mound of earth and one can see amongst them here and there a black Negro head, a brown Hindu face or the olive countenance with slanting eyes of a 'child of the sun.' Elsewhere in a group a Swedish sailor away from his whaleship, a Norwegian reindeer herdsman, a gaucho from La Plata, a Creole from Malabar or Mozambique and many others sit together. They do not debate matters literary, fine arts or politics. They amuse themselves with conversation about their countries of origin and its habits, and describe events they have experienced, because every one crossed many lands and many a sea before arriving in Australia.

Yet both sets of goldfields were far enough inland to make transportation difficult and expensive. As in other rushes, the appetite for luxuries grew stronger with the proximity of potential wealth. Splendour and squalor coexisted. Human relations and the landscape both acquired a thin coating of cosmopolitanism that did little to fool even the casual observer. For all the similarities, however, the Victoria goldfields differed substantially from those to the north. For one thing, they were less peaceful. Partly because of this, though partly for unrelated reasons, the forces of authority were much stricter. It was at Ballarat, the best-remembered of the Australian gold rush towns, that the confrontation between the miners and the colonial administration erupted in battle.

The English novelist Anthony Trollope, who later reported on the South African gold rush, toured Australia twice, first to collect material for his travel book *Australia and New Zealand* (1873) and then, in 1875, to report for the Liverpool *Mercury*. On the latter occasion he wrote of Victoria:

Tales are still current in the colony of the almost fabulous prices which were paid for the means of subsistence at the diggings. There were no roads, and 100 a ton was the common price of conveyance for goods over 60 or 70 miles of ground. The digger who would not walk the distance paid 10 to be carried, and the man who had the money paid it willingly, in order that his hands might be in the gullies a day or two sooner. And we remember in England with what avidity goods were sent out to the new and precious land. A merchant who had aught to dispose of could not send it quickly enough to Melbourne, in order that it might be bought at ten times its ordinary value by men who in the course of a few weeks had been taught to disregard all question of price by the maddening acquisition of pure gold. And thus not only miners came, but also the miners' wants. Houses were built, and banks were opened – and with the banks, schools and churches. And thus a colony was formed, owing its existence almost as much to those who failed as to those who succeeded. Whether it be good or bad to go into the gold trade may be doubted, but there can be no doubt that it is a fine thing to belong to a gold colony.[4]

Such nostalgia set in as quickly as rigor mortis. At the time, Victoria was brawling, uncomfortable, and often dangerous. The danger stemmed partly from the fact that Van Diemen's Land (as Tasmania was called), which was said to house the most unruly convicts, was so close. By one estimate, three thousand convicts escaped across the Bass Strait to the various gold camps, some of them becoming bushrangers. In a land where sheep stealing, like murder, was a capital offence, they were dealt with severely if captured and thus had great incentive to remain at large. Of the Vandiemonians, sometimes playfully known as Demons, who took to the diggings, one was accidentally drowned while being ducked by a gang of miners for some transgression. At Bendigo, another killed his partner and tried to make off with his gold as well as his own. He was taken out and lynched. But the practice was not common, and its foreign nature was indicated by its name – Yankee justice. The great tension, in fact, was between the miners and the authorities over the question of licences and fees.

Bendigo, Ballarat, and Mount Alexander, the three most famous of the many gold towns in Victoria, all sprang up around discoveries related to sheep ranching. Mount Alexander got its start, for instance, when gold was found beneath a sheep pen. In the familiar pattern, it quickly became a cluster of tents, then a community of lean-tos, and finally a town of wooden false-fronted buildings. Bendigo, named in honour of an English boxer, resulted from a chance find near a hut used by shepherds. Little time was required for it to become a complete town, the administrative and supply centre of a goldfield eleven miles wide where twenty thousand miners laboured. Ballarat, from the point of view of the miners, had the worst reputation for strict enforcement of the mining regulations, thanks to the presence of a particular official, Commissioner David Armstrong; but miners everywhere felt that the odds were against them.

In Victoria, a mining licence cost thirty shillings a month, substantially more than in New South Wales, and for that payment a miner received the right to work a plot that measured only twelve feet a side. The goldfields were thus a tightly patterned patchwork set absurdly against a backdrop of vast open spaces. Being shoulder to shoulder with other men's claims made a miner more respectful of property rights, his own and those of his neighbours, even as it increased his frustration and shortened his temper. The close quarters did nothing to improve public health, and the authorities had to take control of the water supply in order to slow the spread of dysentery. Miners stood to lose their claims if they went for longer than twenty-four hours without performing some work on them.

Along with its hotel, its restaurant, and its brothel, each camp had a tent for the local gold commissioner, who was charged with licensing the miners. There was also a judge. The former was in command of the troopers as well as the foot patrols that were fixtures of every camp. In a manner anticipating American prohibition agents in the 1920s, the gold commissioners staged surprise raids. Diggers without licences would attempt to make themselves inconspicuous, while those with licences were often annoyed at what seemed arbitrary demands expressed in harsh language. Deep shaft mining was an important feature of the Victoria goldfields, particularly at Ballarat, and miners were especially resentful when summoned from far below simply to display their documents. In 1854, with a financial crisis apparent in the government, the governor, Sir Charles Hotham (1806–55), ordered licence inspections twice a week after totting up the difference between the number of licences that had been issued and the far greater number of miners to be seen working. There was no social lubricant to lessen the friction when, as continued to be the case, authority and the libertarian spirit rubbed against each other – nothing to prevent the sparks that finally became fire. The constables, who were often alcoholic ex-soldiers, received half of the fines, and they soon gained a reputation for greed as well as for settling disputes arbitrarily.

With the chaotic influx of miners, the free gold was quickly depleted. There was perhaps thus a certain self-deprecating irony in the way the miners called themselves fossickers. To fossick was to rummage and, by extension, to seek crumbs where others had already partaken heartily. Yet from time to time, miners did chance on the burial places of monstrously large nuggets (a similar geological phenomenon would be one of the features of the British Columbia gold rush). In Victoria, the freak nuggets were so large that they were given names, such as the Sierra Sands (a nod to California) and the Lady Hotham (a politic gesture towards the governor's wife). So large were they that they seemed to have personalities or at least life stories.

The biggest of all was the Welcome Stranger nugget, unearthed in 1869 at Mount Moliagul and depicted on the obverse of the one-ounce nugget, the Australian bullion coin introduced in 1987. Its name is said to derive from the fact that it was discovered by two miners whose supply cart had become stuck in the mud. At 2,284 ounces it was probably the largest single chunk of gold ever found, and the wagon, once freed, broke down under its weight en route to Melbourne. Fearful of bushrangers, the fossickers buried the prize in the earthen floor of their cabin and built a fire atop it, around which they welcomed strangers to sit. Later, with impurities removed, the Welcome Stranger fetched £9,436. But nuggets of only a few ounces were windfalls even by the standards of gold rush economics.

Far more than in California or the Klondike, where there was drudgery aplenty, mining in Australia meant an enormous amount of hard work. As in North America, the gold was located in and beneath the beds of streams and rivers, but it usually lay at a considerable depth, not merely in the gravel that had accumulated on rocky bottoms. There was, then, a great deal of earth to be moved, but there was less hydraulic power with which to do so, since the riverbeds were dry part of the year. Whether the miners were dealing with gold sprinkled in earth or embedded in quartz, they had to dig. The small size of the claims, too, demanded vertical holes of narrow width. Thus, the goldfields were pocked with active shafts or disused ones that no one had bothered to refill. The latter posed a real danger to people abroad after dark, both innocent strollers and the so-called night fossickers, or gold thieves, whose fate if caught was a severe punishment from a kangaroo court.

Thomas Mason (1829–1922?) was a typical gold seeker from England who had first gone to South Africa and then endured a typhoon in the Indian Ocean to reach Australia. He later recalled having to shout when walking at night in Ballarat to keep from being shot as a thief while at the same time minding the open shafts. He reported one incident in which a pedestrian fell down a hole with quicksand at the bottom and had to be retrieved in pieces by men with harpoons.[5] Every goldfield had an undertaker. He could be seen removing bodies in a wheelbarrow to an area outside the claims which had been designated for use as a cemetery.

A miner began a shaft the way one digs a grave. When he could no longer throw the shovelfuls of earth over the top, he rigged up a windlass to haul dirt out in buckets or leather bags. Here began the need for miners to work together in pairs or in groups of three or four. If the shaft lacked rock or clay all the way down, the sides had to be shored up with boards to prevent collapse. Sandstone gave the best combination of ease and safety, but the miners knew how to proceed regardless of the subsurface conditions. Experienced fossickers could read the various soils and clays like a haruspex, using the changes to foretell the presence of gold and its quality. Many of the shafts reached a depth of two hundred feet, especially when quartz was involved, as was commonly the case at Ballarat. It was at this point that corporate behaviour became necessary, for shafts could be flooded out by underground streams, and substantial capital was required to bring English-made steam pumps to such a remote corner of the world.

Once the earth was thrown or winched up, it had to be washed for traces of fine gold. This procedure made use of all the various devices perfected in California, such as the sluice, the cradle, the rocker, and the long tom – in short, any method by which the dirt and gravel could be agitated so that the heavier gold became separated from the rest. If water power was the key to this, the gravel might be left in great heaps until the rainy season. High-quality rock brought up from the depths had to be crushed and the gold taken out with quicksilver. A miner who faced the problem of how to process enormous quantities of dirt at rather poor diggings could resort to a puddling machine, by which a horse, walking a circular path, dragged wooden rakes through a pit, whose contents were constantly being changed by a controlled flow of water. A proper puddling operation required four men, so here was another force drawing the miners inexorably towards a corporate culture at the same time as the government was grating on their freedom of movement.

In serious ways and in trivialities, the Australian miners and their communities were far different from those of California, despite what the sequential nature of the two great rushes might suggest. Old photographs and even the more fanciful engravings of the period betray the forty-niners' preference for plain canvas trousers, flannel shirts, and stiff broad-brimmed hats. One sees a different Australian style in the famous lithographs of S.T. Gill (1818–80). The Polish diarist quoted earlier, a former soldier named Seweryn Korzelinski (1802–76) who worked at Bendigo and various other sites, itemized the components: 'Round hats with wide brims, blue striped shirts, moleskin trousers, shoes or boots heavily soled with metal studs. Long beards indicate that either there is no time to take care of one's appearance, or that no one cares about it.' The fossicker might carry a knife in his belt, but more as a tool than a weapon, and typically he also had a tinder bag, a tobacco pouch, and a pipe, as well as a bluestone to rub on wounds and scratches. At the height of summer, some wore veils as a defence against the mosquitoes that were complained of by so many early diarists. Korzelinski observed that even Australians who were not miners but 'future miners, merchants, drivers, farmers and even possibly a future member of the Australian parliament' all affected the miner's style of dress: 'Overcoats, frockcoats and surtouts' disappeared, and egalitarianism became the fashion.

As for food, the information is by no means complete, but there are enough data to suggest that compared with the forty-niners, the diggers had a high-energy, high-protein diet, although it was also rather deficient in fresh vegetables and dairy products. Accounts tell of either steak or mutton, fried in fat, at every meal, with plenty of bread. The billycan might be kept boiling throughout the day. There was a preference in some circles for coffee, sometimes called cafio, at breakfast, but tea of course was standard with other meals. Plum pudding was a special treat on the Sabbath. One would probably not be far wrong in guessing that the beef and mutton were in sufficient quantities to provide the four thousand or so calories one would need each day to undertake long hours of heavy labour. It is certainly true that each gold camp had a butcher's shop.

In California many immigrant entrepreneurs, as well as the merchants based permanently in San Francisco, had been able to supply the camps with whatever their residents needed or desired. Australia offered much the same opportunity, and many a fortune was made. In 1854 Freeman Cobb (1830–78), an American, started Cobb & Co., which soon became one of the world's largest coach lines, with a name that has been at least as famous in Australian history as Wells Fargo's is in American lore. Cobb's secret was that he imported from America the Concord coach – the stagecoach of Hollywood westerns – with its carriage slung on leather straps rather than set on iron springs.

As in California, Jews and other minorities often found the freedom they sought to create new businesses within the framework of the gold rush economy, though the existing merchants of Melbourne had a much firmer grip on the bush-country market than their San Francisco counterparts had on the California mining districts. Although quite small,

Melbourne was already a well-established and tightly structured community when the rush began. It was the clearing-house through which Australian imports and exports passed. As well as being the centre from which the gold rush was administered – and the desire to regulate, legislate, and direct was essential to Australia's gold rush experience – Melbourne also set the prices for what went in and what came out. Economically as well as socially, this gold rush was more complicated than the previous one.

Those members of the Melbourne and Sydney proprietary class who wanted the gold rush suppressed because they feared it would bring confusion were absolutely correct in their assumption. The goldfields were a powerful suction that drew from the cities many of the people who were needed to make them work – needed all the more at a time when immigrants were disembarking in such phenomenal numbers. Between 1851 and 1854, wages rose between 500 and 700 per cent (though by 1856, when the gold rush was losing its allure, they fell to only half their pinnacle). Despite such jagged inflation, the number of bank loans increased fivefold, and the government eventually had to step in to prevent a possible implosion of the credit system (an early example of such fiscal policy). Internal migration, too, was extraordinarily heavy in the goldfields' favour. So many graziers quit the sheep stations that a shipload of replacements had to be brought from the Scottish Highlands. One of the earliest and most popular of the guidebooks to the goldfields was a fraud, written in London by an author whose knowledge came largely from studying S.T. Gill's sketches, though the deception was not exposed for a hundred and twenty-five years.[6]

As in California, the surge of people and goods caused significant improvements in transportation or at least put new ideas to the test. Geoffrey Blainey, the Australian economic historian, has pointed out that John Towson, an English instrument maker, was official examiner of masters and mates at Liverpool when the port was enjoying added prosperity as a result of the Australian gold rush. It was Towson who produced cheap but reliable chronometers in such numbers that every ship's officer could afford to carry one. This meant that navigators were no longer dependent on the Mercator maps, first introduced in the late sixteenth century, which showed the world in rectangular form, thus distorting the size of land masses and the distances between them, and propagating the notion that a straight line along a parallel of latitude was perforce the shortest route, since it was the easiest way to keep a course. Consequently, new, speedier methods came into common usage just as speed was taking on much greater importance.[7] Similarly, clipper ships proved their worth again, not simply because of their sleek design but also because the voyage from Britain to Australia was mostly through great open expanses; clippers did not require coaling stops and were thus free to sail straight down the middle, as it were. The transportation bottlenecks came only after a ship reached Australia.

The roads to the goldfields remained at best inadequate, and though capitalists often spoke of railways to connect the mining districts to the coast, the uncertainty of gold rush economics prevented any lines from taking shape until mining production had levelled off later in the decade. But the gold rush was the reason that the great river systems, the Murray, the Darling, and the Murrumbidgee, were opened to commercial navigation. Even so, goods continued to be far more expensive than even the great distances involved would justify. Fresh oysters were four shillings a dozen in London. Similar oysters were a shilling each in Melbourne. Had anyone at the diggings had a taste for them, they would have increased in price at a multiple far greater than that reckoned on what could be called an oyster-a-mile formula. The reality was not simply that successful miners were prepared to pay for their luxuries; the reality was that transport costs were so high that the commonplace became luxurious. Inflation was constant. The seasonal consideration was merely that prices increased further when the roads became boggy in the rains. In 1854, the year that the licensing debates came to a boil in the fight at Eureka Stockade, miners in Victoria shipped out £7 million in gold, but what they paid out as consumers included £2 million in cartage charges on incoming goods. The towns whose names sound so glamorous now were expensive places to live. The social costs were just as high.

In time, the Victoria gold rush towns took on a long-overdue solidity if not precisely a stateliness. One sees the sense of permanence today in such places as Ballarat. It is in obvious contrast to its equivalents in California, where the culture based on fast food and instant gratification springs naturally from a heritage of disposable towns – quickly built without plan, abandoned just as quickly when the gold played out, and left to collapse. One reason for the greater strength of the Victoria towns is that their existence was, after all, the outcome of at least a half-formed government development policy and not just the result of caprice. The whole nature of Australia made the act of prospecting as much one of settlement as of mere fortune hunting. Although most diggers were young single men who were able to take risks financially (and politically), it was common for entire families to work at the diggings; and it is revealing that in this altogether less laissez-faire atmosphere than California's there was a need for sham marriages as a middle stage between prostitution and respectable mating.

Travelling through Ballarat, Lord Robert Cecil, the future British prime minister, was surprised to see 'a digger in his jumper and working dress walking arm-in-arm with a woman dressed in the most exaggerated finery, with a parasol of blue damask silk that would have seemed gorgeous in Hyde Park.' The woman was what was known locally as a temporary wife, a partner in a mutually parasitic relationship that stopped rather short of symbiosis. Such a woman would extend sexual favours to her grubby companion only for so long as his claim was lucrative and she could enjoy a portion of its wealth. Then she would strike another arrangement elsewhere. It is easy to condemn the practice on one ground or another, but it is usually misleading to superimpose one's own moral environment on historical settings. No doubt gold rushes, which in many ways crystallized the concerns of the nineteenth century, exploited women. Yet from the perspective of today's feminism it is also true that in gold rushes prostitutes were most often free entrepreneurs, not the pawns of pimps or other overseers as tended to be the case in the established cities. In this respect as well as others, being a temporary wife was certainly an improvement on being a simple whore (at least her state *was* temporary).

The government tried to maintain a ratio of six to one in the number of single male immigrants to single female ones, but the policy was difficult to administer and had uncertain advantages, so it was dropped. That many young, usually lower-middle-class Englishwomen came from home in the hope of finding a husband (a permanent one) seems to be well documented, as are the many cases of women who settled instead for domestic service or even became, for a time at least, wards of the colony with beds in the Melbourne settlement houses that had been opened for that purpose. But the organized migration of disadvantaged Englishwomen (in anticipation of the Barnardo Boys of a later time and different place) certainly had its social function. Such good works made a secular saint of Mrs Caroline Chisholm (1808–77), whose career points up the underside of the Australian gold rush but illustrates the good-heartedness that responded to poor conditions. The Reverend George Henry Backhaus (1811–82) is another much-admired figure. He was a Prussian-born priest who had come to Australia in 1846 and looked after the pastoral care of German Catholics in Sydney before moving to Adelaide to work with the poor. When the gold rush broke in 1851, so many people left South Australia for New South Wales and the future Victoria that Backhaus's bishop suggested he return to Sydney. But Backhaus begged to go to the goldfields instead, and he celebrated the first mass in the Bendigo region – in the gold commissioner's tent. He travelled the various camps, tending to the medical as well as spiritual needs of the diggers, and engaging in both social reform and anonymous charity. The careers of both these people represent an advance on California in the gold rush evolutionary chart.

Entertainment in the camps was both individual and organized. Possibly even more so than in California, music was to be heard everywhere in the off-hours. It was in fact the principal cultural activity, and it often came in national costume. There were German singing societies, orchestras, and opera groups, and at least one all-Czech band. The skirl of the pipes could sometimes be heard as a homesick Scot in appropriate dress wended his way among the heaps of detritus. Professional entertainers rightly sensed a profitable market in the goldfields. Black minstrel shows from America are reported to have been seen in Bendigo and were no doubt staged elsewhere. Then there was Edwin Booth (1833–93), the famous American actor who came from a family of the same (his father was Junius Brutus Booth, a legendary tragedian, and his brother was John Wilkes Booth, who in 1865 assassinated Abraham Lincoln). In 1852 Edwin Booth went to San Francisco, where he became a star, and two years later he embarked on an extended tour of Sydney and Melbourne. His success was rapturous, though reviewers complained of his American accent. In later years, when referring to his alcoholism, he often cited 'the accumulated vices I had acquired in the wilds of California and Australia.' A similar inclination to follow the gold rush crusade was a strong force in the career of Lola Montez (1818–61), the actress, dancer, and courtesan whose beaux included Franz Liszt, Alexandre Dumas *père*, and Ludwig I of Bavaria. Her 'spider dance' has come down in legend as having been nothing less than concupiscent. She was a star of the California goldfields before repeating her success in the Australian ones. When the editor of the *Ballarat Times* pronounced on her immorality during the 1856 season there, she responded with a horsewhip.

There are no statistics to settle the question, but a reading of many contemporary accounts suggests that drinking was more of a social ill in 'the Eldorado of the antipodes' than it was in California. But this may be an illusion caused partly by the fact that in Australia, too, alcohol became a source of tension between the miners and administrators. The licence fees for drinking places were set very high: £100 per annum in Victoria. Yet when the tide of events was running in a publican's favour, he could pay the fee from a few weeks' receipts and actually finance construction of his building as well. But the uncertain on-again off-again cycle of boom towns must have added to the general anti-authoritarian streak that resulted in so much illegal activity in this area. Every mining camp seems to have had its sly grog-shops, where the spirits no doubt tasted all the sweeter for their illicit nature. Fines and terms in jail did little to keep these establishments from multiplying. It is tempting to see the spread of them as further evidence of the growing cycle of defiance and repression.

All manner of people went to the diggings for all sorts of reasons. James Bonwick (1817–1906) was an educator who arrived in Victoria in 1852 in the hope of striking it sufficiently rich to continue construction of a school he was building in Adelaide. The dream proved elusive, but he founded *The Australian Gold Digger's Monthly Magazine and Colonial Family Visitor* and thus set in motion a long and fruitful career as publisher, editor, writer, and archivist. Philip Francis Adams (1828–1901) was an English surveyor who worked in Canada and the United States, was unsuccessful at prospecting in California, and so in 1854 turned up in Sydney. In time, he became an authority on viticulture and wine making as well as a notable astronomer. As with other rushes, there were also those who went to escape persecution and perhaps prosecution too. They came from every direction. As usual, the acquisitive nature of the enterprise, the close quarters, and the polyglot composition of the mob made discord inevitable.

The Aborigines had no serious trouble with the diggers (a marked contrast to the later gold rush in Queensland), for by the 1850s they had already been driven away from the settled areas of New South Wales and Victoria and their numbers had been reduced by smallpox and influenza. When contact did take place, the whites viewed them with contempt. This attitude went back to Captain Cook, who believed that since the Aborigines did not farm, they did not truly use the land. In fact, they were so tied to the land that they began to perish without the full use of it. In the New Zealand gold rush that followed, there was at least a general awareness in government circles that the Maori (who, by contrast, were an agricultural people) should be compensated for the mass act of trespass.[8] But in neither case was there the fear that if the rush was not contained, the indigenous people would rise up – as happened in the United States two decades later with the Sioux in the Dakota Territory.

Perhaps the most numerous diggers, aside from British immigrants and the Australians themselves, were the Germans – a fact that has not made a lasting impression in the popular imagination. This is possibly because of anti-German sentiment in the twentieth century. The most conspicuous of all the groups were certainly the Americans, whose brashness and lack of culture were as always a subject of fascination and distaste. Korzelinski tells the story of an American miner who, when challenged to name the two greatest citizens his country had produced so far, answered without hesitation and in deadly earnest, 'General Jackson and Jesus Christ.'

As with California and the forty-niners, one can sometimes turn to popular fiction about the gold rush to get an honest insight into how the participants would have spoken and behaved. The best-known examples of the genre are by Thomas Alexander Brown (1826–1915), who under the name Rolf Bolderwood wrote *The Miner's Right, The Squatter's Dream*, and *The Ghost Camp*, among other books. Particularly useful is his *Robbery under Arms*, which contains a scene in which two Australians had to evade the police by passing themselves off as Americans. The impersonation was apparently simple enough because of the exaggeration inherent in genuine Americans, who 'were all such swells, with their silk sashes, bowie knives, and broad-leafed "full-square" hats, that lots of young native fellows took a pride in copying them, and could walk and talk and guess and calculate wonderfully well considering.'

As usual, there was systematic hatred of the Chinese, who nonetheless came in great numbers, most of them working quietly and productively. A liberal Australian settler writing in a British journal expressed wonder at the way 'the pride and prejudice of "race" displayed itself against the Chinese emigrants to an extreme degree.' He added: 'They had, in fact, to struggle with persecutions sufficient to have daunted the energies and depressed the spirit of any less adventurous people; and though much of this hostile reception may have been provoked by their native obstinacy, it undoubtedly derived its strongest impulse from the very general aversion of the colonists to a Tartar influx.'[9]

It is believed that Louis Ah Mouy (1826–1918), through a letter to his brother in Canton, provided a great deal of the initial impetus for Chinese immigration. Certainly, Ah Mouy became the dominant force in the Chinese community in Victoria. He was working as a builder when the rush broke, and made himself rich through a claim in the Yea region.

He used the money to establish himself as a merchant and to speculate in mining ventures not only there but also at Ballarat, Mount Buffalo, Bright, and Walhalla, using Chinese miners. Later, he opened up gold properties in Malaya as well.

By June 1854 there were two thousand Chinese working the Victoria goldfields, and the following month bloody anti-Chinese rioting erupted at Bendigo. The next year there were seventeen thousand Chinese – more than in California – and Victoria imposed an entry tax of ten pounds a head, the first in a long code of anti-Chinese legislation, which the people for whom it was intended evaded whenever possible. The common method was to avoid Melbourne, come ashore in South Australia, and then quietly move cross-country. It was while they were engaged in this secret migration that a group of Chinese miners discovered the gold-fields at Ararat. In 1857 there was further rioting, this time at Buckland. The fact that both riots took place on 4 July lends some weight to the view that Americans were the main instigators. They somehow believed that their Americanism required them to take action against Asians and that this rule obtained wherever they happened to be.

The most persistent threat of violence and mayhem was external. It came from bushrangers, the escaped convicts and other misfits who 'bailed up' travellers and skirmished with the troopers. Ned Kelly, whose day came later, at the end of that period of lawlessness, is merely the most storied of hundreds. People were loath to travel alone. The tales Korzelinski was told were typical: 'Sometimes they tie their victim to a tree and leave him to the ants, mosquitoes and hunger. Very rarely is the unfortunate victim found in time, more often one finds a skeleton tied to a tree. Hardly anyone travels with gold.'

The bigger the bushranger gangs, the greater was their temerity. The worst single incident came in July 1853 when a large party descended on a gold shipment that was being sent to Melbourne by a private company. More than two hundred ounces were stolen and four of the six guards killed. Only three of the outlaws were ever apprehended, and they were promptly hanged. Most shipping of gold was done by the government, and for once the miners had reason to be thankful for the troopers' presence. The diggers brought their gold to the commissioner's tent for safekeeping. When a sufficient amount had accumulated, the gold was sent to Melbourne in a two-wheeled cart drawn by four horses and surrounded by heavily armed troopers. The departures were irregular, but bushrangers sometimes managed to learn the schedule nonetheless, no doubt through the use of paid spies. It was for this reason that the convoys always proceeded at the gallop and thus were forced to make frequent stops at places that could be defended easily. Even in Melbourne there was at least the possibility of a daring robbery before the precious cargo could be placed in bank vaults. There is still at least one old coaching stop in Melbourne, Mac's Hotel in Franklin Street (1853), with special quarters that were built for the gold escort.

By such means a vast amount of gold, weighing as much as one million pounds in all, was transported from the Victoria fields between 1851 and 1856. Once in Melbourne, most of it was loaded aboard ship, for it was not until 1869 that the first unsuccessful attempt at an Australian mint was begun. The traffic gave shipowners, who traditionally could find little backloading after delivering their passengers safely in Australia, a new and profitable business, though from time to time a loss would rebound heavily on the assurance companies. In 1859 the steamship *Royal Charter*, en route to Britain, was wrecked off Anglesey in the Irish Sea with Australian gold listed at £375,000. More than four hundred persons drowned, making this one of the worst peacetime disasters at sea to that date. In recent years, the wreck has been located by British marine archaeologists. In 1866, when gold shipments from Australia had slackened considerably, the steamer *General Grant*, en route from Melbourne to London with 2,567 ounces of gold and many of the men who had dug it, went down in the Auckland Islands, south of New Zealand. Only fourteen people survived. Divers continue to search for its remains.

The trouble that had been building in Victoria might have boiled over at any number of spots, but Ballarat was to be the place. To Sir Charles Hotham and his officials, the result must have seemed puzzling at times. The miners from Britain, whose level of education was remarkably high, were undoubtedly shot through with Chartism. And it seems not unlikely that some of the Germans had been red forty-eighters and were electrified with a radical ideology of their own. But had not Hotham at least been uniform in his treatment of the miners? The licence fees, which were thought confiscatory, were the same for everyone in the colony. If the price was often too high for individual diggers engaging in what after all was a highly speculative venture, it was not at all too high for a group of miners working together – a reminder that bureaucracy is more comfortable dealing with business than with the public. But the fees were only one point of contention. The cavalier use of troopers to sweep down every few days demanding documents was foreign to the cultural and political traditions of freeborn Englishmen and their descendants (whereas at a gold rush in French Guiana, for example, the tactic might not have been so disruptive). Many miners also resented the fact that diggers did not enjoy the franchise. In this they were probably influenced by the always vocal Americans, with their harping on the virtues of corn-pone democracy. Yet the fact remained that other British colonies, such as the Province of Canada, had won responsible government, though even a Whig might point out that Victoria and New South Wales were not yet ready to take on the burden – though, Lord knows, their economies were surging ahead. The miners' difficulty in getting land on which they could settle was another issue. All such factors tended to run together into one cry of protest that was nominally about licences. It was a chant heard almost from the beginning.

In 1851 miners had met at Buninyong to protest the announcement that licences would be introduced. They had been vehement and had vowed to abandon the place unless their appeal was heard. In the event, it was the protest not the camp that was abandoned, when a short time later a major new field was discovered. But the precedent augured ill all the same. During 1853 a group calling itself the Red Ribbon League was formed. It was composed of many nationalities, though Germans were conspicuous in the leadership; it was no doubt they who were influential in constructing the thoroughly anti-British platform, which nonetheless proposed sensible if sweeping reforms, such as a tax on gold at the point of export rather than on miners at the point of production. A boycott of the hated licence fee was begun. Each miner in favour was to decorate his hat with a red ribbon. Whipped into a frenzy, the diggers armed themselves and marched to the commissioner's tent at Bendigo, where they were met by Joseph Panton (1831–1913), an assistant senior commissioner (the bureaucracy was becoming stratified). Drawing deeply from the cool well of civil-service charm, he promised them a fair hearing for their views in Melbourne. He was true to his word, and when the three-monthly fee was temporarily reduced from four pounds ten shillings to two pounds, the situation was defused. Panton's attitude should have been a model for his master, Sir Charles, but alas it was not.

Ballarat was the richest camp but also the one whose gold was most difficult to find. The hit-or-miss technique of sinking shafts and cranking windlasses, hoping that one's excavation a few feet in width would magically intersect one of the north-south quartz seams a few inches or a few feet wide, combined the greatest expenditure of time with the maximum of uncertainty as licences ate away at one's capital. Once miners had banded together to share expenses, it was a short step to banding together for other purposes. A wiser man might have seen the danger more clearly, but the governor was arrogant and stubborn, as disdainful of the rabble, whom he tended to regard as savages, as he was inclined to worship his own importance. He weakened the police force when he might have strengthened it – indeed, when the miners wished it improved for their own protection. And while spending lavishly for his personal privacy and comfort, he did away with the gold commissioners' mounts so that they had to make their rounds of the camps on foot.

It was a sensational murder case that sparked the violent confrontation. An ex-convict named James Bentley (died 1873), who kept a hotel in Ballarat, was accused along with his wife Catherine and a partner named John Farrell of murdering a miner, James Scobie. A magistrate with whom Bentley was friendly found him innocent, whereupon the miners ran riot, sacking and burning the hotel. The governor ordered his men to arrest the mob leaders as well as rearresting the three accused, and this brought rebellion to the boil.

The dissidents now called themselves the Ballarat Reform League and drew up a list of demands. Above all, of course, they wanted the licence fee abolished; many burned the hated pieces of paper then and there, and still others vowed never to pay the tribute again. They also insisted on the right to elect representatives to sit in Melbourne. By now, surface mining was rapidly dying out and shaft mining was becoming more complicated and expensive, with the result that many miners were drifting into towns or squatting on vacant Crown land. That they be allowed to purchase such land was the third item on their agenda. In a meeting at Eureka, a gold camp near Ballarat, the miners also petitioned the commissioner to release their colleagues from jail. This was refused.

In one of the most dramatic turning points in Australian history, some 150 miners threw up a circular barricade of planks and stones, about 300 yards from the site of the present-day reconstruction. They huddled behind it, well armed, flying a homemade flag called the Southern Cross, which is now in the Ballarat Fine Art Gallery. The design, by Captain (or sometimes Lieutenant) Charles Ross, a Canadian from Toronto, about whom virtually nothing else is known, was the basis for the modern Australian flag.

On 3 December 1854, Commissioner Robert Rede (1815–1904) drew up to the rebel fort with 276 constables and troopers on foot. After midnight, when only a few sentries were awake, he reduced the stockade by storm in only twenty minutes. The first volley is thought to have inflicted a withering fire on the miners, many of whom did not have their weapons to hand. Nonetheless, five soldiers were killed. On the miners' side, there were about thirty deaths. Both forces had three times as many wounded as killed. Of course, the defenders accounted for only a tiny percentage of the local miners. Their job at the stockade done, the troopers then repaired to the goldfields and inflicted more casualties on individual miners. Sir Charles Hotham later had the good grace to say that Rede had acted imprudently.

The acknowledged leader of the rebels, Peter Lalor (1827–89), an Irishman, escaped. The handbill circulated to announce the £400 reward for his capture described him in flattering terms: 'whiskers dark brown and shaved under the chin ... good looking, and is a well made man.' Although he lost an arm as a result of wounds incurred in the skirmish, Lalor evaded his trackers until the reward was withdrawn. A look at some of the other miners prominent in the action at the crude stockade shows the diverse nature of the mining community. Ross, the Canadian, was killed in the assault. Frederick Vern, a German, escaped in the confusion. Lalor's lieutenant, an Italian, Raffaello Carboni (1820–75), was captured and published the only substantial first-person account of the engagement, writing in a style so highly individualistic as to be incomprehensible at times. Capture also awaited C.D. Ferguson, an American forty-niner. In fact, a number of Americans were captured, but the U.S. consul in Melbourne exerted his considerable influence to have them freed – all except John Joseph, the only African American among them. He stood trial for treason with a small group of others, but all were acquitted, confirming the decision already handed down in the court of public opinion.

Before long, all the miners' original grievances were satisfied. The licence was replaced by a so-called miner's right, which was available for a mere one pound per annum, and legislation was introduced to allow those dispossessed by the rapidly waning gold rush to buy land for settlement. More importantly, the fiasco at Eureka Stockade contributed immeasurably to the inevitable introduction of responsible government in 1855 (when, lo, Peter Lalor was elected to represent Ballarat in the first legislative council – and soon came to seem a conservative). For generations afterwards, however, Australians were reluctant to admit the importance of the fighting. Their hesitation no doubt came from complicated emotions and cultural factors.

After about 1860 there were no longer diggers but only miners employed by the mining companies. The conservative atmosphere of corporate mining caused people to shrink from the memory of what had happened at the stockade. One of Harry Lawson's short stories synthesizes the mood perfectly when it describes two old friends who take a walk in the dark after dinner and allude *sotto voce* to the events of twenty years earlier. From being only a whispered memory, the Eureka Stockade was taken up as a radical symbol by a later generation. In the 1890s the striking shearers flew the Southern Cross; later, so did the Builders Labourers' Federation, one of the most militant Australian unions; and during the Depression, the young people's wing of the Communist Party of Australia was named the Eureka Youth League. It was only as recently as 1947, with the feature film *Eureka Stockade*, that the event completed its entry into the cultural mainstream.

Much can be learned by comparing the legacy of the California gold rush with that of the Victoria rush. Americans have always considered theirs a happy event, associated with material success; in the national memory it is accompanied by strumming on a tenor banjo. Australians have been less sanguine and in any case more subtle. For instance, the important gold rush images in Australian painting, such as *The Prospector* by Julian Ashton (Art Gallery of New South Wales), convey a mood of defeat rather than triumph. The gold rush is also the source of that curious tribal rite known as Australian Rules football, many of whose terms, including 'shepherding,' 'giving a lead,' and 'pockets,' derive from early mining. Diggers carried the game to South Australia, Western Australia, Tasmania, and New Zealand, but it is still not played in New South Wales. In California there is a professional football team called the Forty-niners, but that is the extent of the inheritance.

Yet Americans who have been aware of the Eureka uprising have often regarded it as an event that proved the righteousness of what they consider the American ideal. Mark Twain set the pattern when he called it 'the finest thing in Australasian history':

It was a revolution – small in size, but great politically; it was a strike for liberty, a struggle for a principle, a stand against injustice and oppression. It was the Barons and John, over again; it was Hampden and Ship-Money; it was Concord and Lexington; small beginnings, all of them, but all of them great in political results, all of them epoch-making. It is another instance of a victory won by a lost battle. It adds an honorable page to history; the people know it and are proud of it. They keep green the memory of the men who fell at the Eureka Stockade, and Peter Lalor has his monument.[10]

Whatever else it did, the event marked the effective end of the Victoria gold rush. Gold camps, so called because they were understood to be only temporary communities, had grown into established towns and cities. Gold mining had become an Australian industry, and the crusade would eventually gather its forces and push off once more. But a part of the popular imagination of the British Empire, and to some extent Europe and America, would forever after respond to certain catchwords with a vivid and not altogether inaccurate picture of what the goldfields had been like. The plot of the Sherlock Holmes story 'The Bascombe Valley Mystery' revolves around a respectable Englishman whose past comes back to haunt him when it is revealed that he was once known as Black Jack of Ballarat – a bushranger. To Conan Doyle's middle-class audience of 1891, the reference was perfectly plausible yet deliciously romantic. In any case, it played on their collective memory and imagination.

3

To the Ends of the Empire

While the California gold rush was going at full speed in 1849, a Scot named James (later Sir James) Douglas, far up the Pacific coast in what would one day be part of Canada, was viewing with alarm the entire vulgar display even as he encouraged the local Native people to search for gold that would make him and his masters rich. His masters were the governors of the Hudson's Bay Company, which at the time controlled virtually all of what is now western Canada as well as much of the North. As the company's chief factor in the region, Douglas (1803–77) wielded enormous power.

In 1821 the Hudson's Bay Company had amalgamated with its great rival, the North West Company, thus gaining access to the fur trade west of the mountains. But this brought it into conflict with the Americans because of the indistinct nature of the boundary. At the time, the only Europeans in the Oregon Territory south of the forty-ninth parallel were Hudson's Bay Company men, who regarded it as their fur-trading preserve; but by the 1840s the scene had changed dramatically. American attention had focused on the area, and settlers were pouring in. Some even asserted that the entire coast, all the way to the Alaska border north of the fifty-fourth parallel, rightfully belonged to the United States. 'Fifty-four forty, or fight!' was their cry.

The fact that the Americans were on the verge of declaring war on Britain for the third time in seventy years had a great impact on the California rush. Not only did it increase the number of settlers on the coast – and hence the number of potential gold seekers – but the Oregon Trail which brought them west could also be used by other argonauts. When the Oregon Boundary Dispute was settled in the United States' favour, giving it all the mainland up to the forty-ninth parallel, even more settlers arrived. There were thus thousands of energetic individualists living in the area when a decade later gold was discovered in British territory to the north.

In the boundary treaty the United States did at least agree to give up all claims to Vancouver Island, where James Douglas had established a fur-trading post a few years earlier. Britain hastened to consolidate its position by arranging for a colony to be established on the island in 1849 under the auspices of the Hudson's Bay Company. As the senior company official on the coast, Douglas expected to be made governor and was chagrined when the Colonial Office chose someone else, a certain Richard Blanshard (1817–94). But Douglas soon succeeded in making life so difficult for the new governor that Blanshard's health gave way and he resigned, leaving the way open for his rival. Thus, in 1851, Douglas received the coveted appointment while retaining his position with the Hudson's Bay Company.

Douglas was a businessman, however, not a colonizer. A volatile – indeed, explosive – personality, he was autocratic at a more basic level than even his powerful position might warrant. He did little to encourage settlement, and when settlers did begin to arrive he took his time about establishing a legislative assembly, finally getting around to it only after much prompting from Britain. Even then, Douglas made sure that the legislature had little power. Too much democracy was dangerous – it smacked of American republicanism and disdain for the rule of law. Douglas had done his best to keep the news from the Americans when in 1850 gold was discovered on the Queen Charlotte Islands. Nonetheless, word of the find spread, with predictable results. In 1852 – when, significantly, California was losing momentum – five hundred American prospectors headed for the Queen Charlottes. But they were outmanoeuvred by Douglas, who had already sent some company men there to work the only known vein.

In the unstoppable course of events, gold was discovered on the mainland opposite Vancouver Island in 1856, near what became Lytton, where the Thompson River enters the Fraser. The company again attempted to impose silence, but with equal lack of success. Soon Americans were moving north across the border, much to the dismay of Douglas – and also to the fury of the Native people, who resented the presence of these rival gold seekers on their territory. Clearly, quick action had to be taken, both to keep the peace and to prevent yet another takeover by the Americans. The loss of the Oregon Territory was still fresh in everyone's mind.

Although Douglas was the senior fur trader on the mainland, he did not have the authority to issue edicts, but as governor of Vancouver Island he was the nearest representative of the British government. He sent dire warnings to London in 1857 and later in the year issued a proclamation asserting the Crown's control over mineral rights and requiring all miners to pay a licence fee. He was quick to justify this action, explaining:

My authority for issuing that proclamation, seeing that it refers to certain districts of continental [North] America, which are not strictly speaking, within the Jurisdiction of this Government may perhaps, be called in question; but I trust that the motives which have influenced me on this occasion and the fact of my being invested with the authority over the premises of the Hudson's Bay Company, and the only authority commissioned by Her Majesty within reach, will plead my excuse. Moreover, should Her Majesty's Government not deem it advisable to enforce the rights of the Crown, as set forth in the proclamation, it may be allowed to fall to the ground and become a mere dead letter.

What HMG did in fact deem advisable was to invalidate the company's monopoly on the mainland and make the region a British colony. It was named British Columbia at the suggestion of the Queen herself, who pointed out that the Americans had appropriated the word Columbia, in unmodified form, to represent themselves in patriotic song and verse. Douglas was to be governor of this colony as well as of Vancouver Island on condition that he resign from the Hudson's Bay Company, which he agreed to do.

The act creating British Columbia was passed in August 1858, by which time at least ten thousand miners were panning the lower reaches of the Fraser and inching northward. Ironically, it was Douglas who had opened the floodgates. In California, the purchase and coining of gold had been left to a ragtag collection of private interests until a federal mint was established in San Francisco in 1854; but in Victoria, Douglas bought much of the gold from the miners, just as he had bought small amounts from the Native people, and in February 1858 he shipped 800 ounces to the San Francisco mint for refining. It was the ship's arrival there, with its irrefutable proof of wealth, that touched off the big rush. Douglas learned the lesson and later attempted to mint gold coin at British Columbia's infant capital, New Westminster – without, of course, official sanction to do so – but by then the hordes had come.

Most argonauts came by sea, for a vicious war was being fought between the U.S. Army and the Native people just south of the border. The first ship from San Francisco arrived at Victoria in April 1858 – much to the surprise of the tiny settlement's inhabitants, who were all in their Sunday best, having just come out of church. They were relieved to find that the four hundred or so roughly clad adventurers did not desire to stay; they simply wanted to fill up with food and other supplies before crossing the Strait of Georgia. By the end of June, seventeen hundred miners had passed through Victoria; by the end of December, thirty thousand. The original fort on the tip of Vancouver Island was surrounded with tents as the hopefuls jostled one another for accommodation – always a premium commodity, despite the attendant building boom which transformed Fort Victoria the trading post into Victoria the city. Over a period of one and a half months, one witness counted 225 new commercial buildings, which were by no means all the jerry-rigged sort associated with San Francisco's first flurry of excitement. The surveying business was especially brisk. Before the year was out, the community had acquired many of the basic amenities. The first book to be published in British Columbia was an 1858 work by an Anglo-Frenchman who had been in the California rush. He was Alfred Waddington, the author of *The Fraser Mines Vindicated. Or, The History of Four Months*. Not far behind in its timing was *The New Eldorado, or British Columbia*, by Kinahan Cornwallis, which included a map that showed with uncanny accuracy the places in the interior where gold was later discovered.

The miners who clogged the street of Victoria faced logistical problems somewhat like those faced by the Klondike gold seekers two generations later. Boat building was one of the skills they had to cultivate in order to transport themselves two hundred and fifty miles across the strait to the mouth of the Fraser. They launched homespun craft of all kinds, many of which were swamped and their owners drowned. The first big strike was just downstream from Yale, an old trading post about one hundred and fifty miles farther up the Fraser. In all, 100,000 ounces were taken out the first year. It was fine flourlike gold, found in rock that first had to be crushed to make concentrate, which was then cleaned with nitric acid. The metal was recovered from the concentrate through the use of mercury (which draws heavier substances such as gold and rejects gravel). Beyond Yale, navigation was dangerous when possible at all, because of the hundred-mile canyon, whose boiling waters splintered small boats against the steep rock walls. During the summer of 1858, more than ten thousand miners were encamped south of the canyon, panning and washing in the short stretch of country between Fort Langley and Yale. Not until the autumn, when the river was lower, did the canyon become even a manageable natural obstacle, thus opening the way to the still more lucrative goldfields that everyone seemed to know existed.

For the men at the diggings, life in British Columbia was substantially different from life in California. The sheer inhospitality of the place was one guarantee of that. Whereas it might have cost two hundred dollars and taken one hundred and fifty days to travel from New York to San Francisco, it cost perhaps thirty-five dollars and took only fifty-five or sixty days to travel from San Francisco to Victoria, a piddling distance by comparison when plotted on a globe or an atlas. On arrival, the gold seekers found the country harsher than California, with mountain snows five feet deep, unnavigable rivers, treacherous canyons, and forests unbroken by roads. Although the terrain was better suited to sustaining life, it was also more threatening to life. The sense of remoteness increased as the miners (the term 'argonaut' had faded) pushed north to places whose names would form an unforgettable litany: Lillooet, Williams Lake, Soda Creek, Alexandria, Quesnel, Barkerville. The names were more deeply indebted to the Native languages than those in California and were not quite so suggestive of violence and adventure as, for instance, Hangtown and Poker Flats. But then there was a sternly benevolent presence, a representative of the Great White Mother from across the oceans, to oversee the rush.

It was partly the remoteness and partly the competition from Australia – and also from Nevada, where the Comstock Lode was attracting a wide range of silver miners – that caused the Fraser River gold rush to be at first so thoroughly American, rather than being a magnet for various nationalities as California had been. This is what alarmed both Governor Douglas and the Crown. At one point, British warships had to cruise conspicuously near Vancouver Island to dampen U.S. territorial ambitions. Meanwhile, Douglas moved to prevent the lawlessness he had heard about in California and to keep the economic benefits of the rush from being drained away across the border. He ruled that Americans could not own land in British Columbia. To prevent them from resorting to vigilante justice or indeed from establishing their own mining code, he imported a Scottish judge from England, Matthew Baillie Begbie (1819–94). He also established a militia regiment to be used as a constabulary. It was his idea, a combination of parsimoniousness and good intentions, to recruit the outfit from the many escaped American slaves and free blacks who were coming north for the rush. But James

Douglas's Coloured Regiment, as it was called, was unsuccessful, because white Americans refused to respect the blacks' authority. It seems that one could come closer to restraining the naturally rowdy Americans by trying to second-guess American responses. Thus, an additional British-trained magistrate, Peter O'Reilly, later talked to miners at Wild Horse Creek in their own vernacular: 'Boys, I am here to keep order and to administer the law. Those who don't want law and order can git, but those who stay within the camp, remember on which side of the line the camp is. For boys, if there is shooting in Kootenay, there will be hanging in Kootenay.'[1]

Much of the tension was racially inspired. The Salish people, on whose land the gold was discovered, greatly resented the white intruders, who responded by proffering liquor and by enslaving the Salish women right under the nose of the Crown. When the Chinese began arriving in great numbers, seeking the least attractive work at half a white labourer's pay, much of the hatred was focused on them, and so began a tradition of Canadian discrimination against Asians, which for duration and virulence was little better than that practised in the United States. Evidence for the point is weak, but the absence of testimony to the contrary piled atop the California precedent suggests that when, late in the decade, many French, Italians, Germans, and Mexicans came to the Cariboo, the effect was to create identifiable communities that were more or less in harmony with one another, in place of two or three powerful racial blocs perpetually at hazard.

A miner named Joseph Haller, writing in German to his family in Pittsburgh, was frank about the peculiarities of the British Columbia rush, noting economic details and commenting on the problem of transportation, which was so much more acute than in California, whence he had come. In autumn 1858 he reported from 'somewhere on the Frazer River' that he was taking out between eight and ten dollars' worth of gold a day. But he added, 'Food is very dear, one pound of flour $1.50, pork $1.75 per lb., beef $1.00 lb., sugar $1.50, beans $1.45, tobacco $6.00. Everything is dear but we hope it will get cheaper. I would like to send $100 to you, but it is impossible; I am too far in the mountains, it will cost more than $50.00 to send it and I would not be sure if you got it or not. I am anxious to hear from you.' A year later he was still mining but taking in only five or six dollars per day, while the cost of sending money to Pittsburgh had doubled; 'and I would have to risk my life,' he said, to carry it to the nearest post office. But consumer prices were falling on the Fraser, and when Haller took the hundred-dollar plunge and went to Victoria to sit out the winter of 1859–60, he found 'everything is cheaper here on Vancouver Island' – pork 25¢ a pound rather than $1.75; tobacco a mere 50¢, not $1.45. Such figures are a powerful indication of how difficult it was to supply camps in the remote stretches, a task usually carried out by mule train. Indeed, Haller gradually withdrew from mining in favour of the teamster business. This led him in turn to operate a hotel, 'a store with grub,' and an establishment called the Red Dog Saloon, fifty-five miles above Lillooet, where he remained permanently and prospered.

In the spring of 1859 a miner named Peter Dunlevy and four fellow Americans struggled up the Fraser as far as the mouth of the Chilcotin River, where a Native (all thought of whom was soon dismissed) led them to the first big strike in the Cariboo region. During the summer, three hundred miners reached Lytton overland, and a little farther upriver they began to find nuggets weighing one-third of an ounce; an experienced miner could wash six ounces a day, it was said. Almost simultaneously, gold was found on high terraces in the semi-arid country north of Lillooet, while still farther up the Fraser, at Fort Alexandria, a Cape Bretoner named Benjamin McDonald found similar high benches with nuggets as large as six ounces. By the spring of 1860, hundreds of miners were working such places, and the first permanent settlement in the area had grown up at Quesnel Forks, complete with the mandatory Chinatown. For in its social and economic organization, gold hunting was suddenly becoming more complicated, a process that matched the more complex geological aspect of this gold rush compared with California's.

It was in summer 1860 that Doc Keithley, an American adventurer, went from the new town of Quesnel Forks to nearby Cariboo Lake where, with another American, J.P. Diller, he made perhaps the richest strike in British Columbia up to that time. From Keithley Creek, as Keithley named the spot unselfconsciously, the same party went north again to the much smaller Cariboo Lake, and in the steep hills that protect it found nuggets of heavily oxidized, or 'sunburnt,' gold. The gold actually littered the ground, a surreal confirmation of what Spanish conquistadors and raggedy European immigrants from the cattle boats had always imagined. This place they called Antler Creek. But when some of the party returned to Keithley Creek for building supplies, word got out. By spring, Antler Creek had its own sawmill, which had already turned out as many as seventy buildings: hotels, saloons, gambling halls. What ensued was a new phenomenon, an actual staking rush, with claimants frantically trying to file legal title to certain small patches of ground instead of simply digging in, each man for himself. Ironically, Keithley's discovery of the free gold lying on the surface was actually an indication that erosion had deposited it there from great veins located higher up. The lone independent placer miner with his one dish and one shovel would continue to exist, but it was clear that the big money could be made only through joint effort. Capital had entered the picture.

The focus of mining in the British Columbia interior shifted from the gravel and sediment found in stream-beds to the hills above the rivers, where layers of gold-bearing rock were spread like secret jam. Hydraulic methods were necessary. Canvas hoses were fitted with brass nozzles and fed with water pumped from below to wash away the overburden. Then the likely layer of mud and rock was flushed down ditches dug for that purpose, ditches that sometimes ran for miles. Ditch digging became a distinct profession, one at which the Chinese proved themselves the most adept (or the least unwilling); in time, ditches were dug in order to be sold to miners working the hills above, as a commodity and form of speculation. When water power could not expose the hiding place of gold, vertical shafts were dug – and then had to be kept dry by means of waterwheels two storeys high, constantly pumping. It was a new type of gold mining and it took money. Companies were formed, shares subscribed, middlemen born.

In February 1861, the earliest possible moment at which to attempt working the thawing ground, still another American, William (Dutch Bill) Dietz, hit upon a previously unreported condition at what he, in exercise of his God-given right, would christen Williams Creek. Eight feet below the surface he hit bedrock and bluish clay; he gave up. The claim passed to others, who dug a further ten feet and found gold. Clearly, such mining called for knowledge of natural science as well as of capital markets. In 1862 a third person approached the spot. He was a Bavarian, Ned Stout, who had taken part in the California and Comstock rushes. He found an entirely new layer of gold-bearing rock below the one already known. On this evidence, Billy Barker, who was destined to be a famous man, was encouraged to dig near by.

Barker is frequently written of as a gold rush archetype, for he made an enormous fortune quickly, squandered it with even greater speed, and died in poverty, ignored by all those to whom he had given money and favours in his reckless desire to celebrate his miraculous life. But he is a useful symbol for the Cariboo in particular, because of his previous history and his method of organization. Barker was English, from Cambridgeshire, but had left to seek gold in California. Finding little, he lingered in the United States until news reached him of the Fraser country in 1858. He lent his name to Barker & Co., a syndicate of himself and seven other miners. It was this company that on 17 August 1862 made the big strike on Williams Creek. Fifty-two feet below the surface the operators discovered greenish-yellow nuggets well worn by some underground water source. They pushed on and at almost eighty feet discovered the vein of bright sharp-edged gold from which the other had come. By 1867 as much as 38,000 ounces of it would be mined. Almost as soon as the strike became known, a settlement called Barkerville began to grow up around the shaft-house. It became a permanent community and a rainbow one – for the first time, Europeans, Australians, and indeed eastern Canadians began to arrive in large numbers, thanks to another and the greatest of Governor Douglas's initiatives, the Cariboo Road.

The problem from Douglas's point of view was how to increase the flow of men and supplies to the mining camps and also keep the benefits from following the line of least resistance down the Okanagan Valley into the United States. Even before resigning from the Hudson's Bay Company, Douglas had been working on a mule trail that was to run from New Westminster to Lillooet, avoiding the Fraser altogether. Five hundred miners, the sort Douglas generally had so little use for and thought devoid of public spirit, volunteered to help build it, even going so far as to pay the company to transport them to the starting point. But the trail, though it avoided the raging Fraser, entailed too many other obstacles and portages to be practical. Even so, those largely unsung forces in Canadian history, the Royal Engineers, were working to improve the road when gold fever spread beyond the lower Fraser into the Cariboo. Thus, in 1862 the sappers began the Cariboo Road, a formidable engineering undertaking indeed, running four hundred miles from Yale to Barkerville, some of it blasted through rock. They completed the two most difficult sections themselves, and the rest was let out to private contractors, who were also encouraged in other roads. (Alfred Waddington, the writer who had beaten everyone else into print with a guidebook, secured a contract to build a road from Bute Inlet on the Pacific, northeastward across the mainland to where the Chilcotin River meets the Fraser, but the plan collapsed after nineteen of his men were killed by members of the Chilcotin Nation.)

The Cariboo Road was completed in 1864 and at once caused the flame of enterprise to flare. One man imported a herd of camels for use as beasts of burden on the trail, but the poor animals were unsuited to the climate and escaped at the earliest opportunity; the last live camel sighting in British Columbia took place in 1896. Even while the road was being built, it helped attract miners and settlers from eastern Canada, the overlanders of countless Canadian schoolbooks. The increase in the number and variety of miners was so great that 1863 was a record year in the Cariboo, with 244,597 ounces mined. And because of the general reduction in the price of supplies and equipment, the road helped keep the mining industry robust even after all the shallow deposits were exhausted. More important to Douglas, it conserved the economic benefits in all the British Columbian camps and towns along the way. Unfortunately, though, he had spent £400,000 on the road without proper authorization and so was eased into retirement in 1864.

Billy Barker met a sadder fate. The place that bore his name was no mere camp or even town but a mining city of several thousand people, including a wealthy class who wintered in Victoria or San Francisco. In quick order it had a library, a theatre, and a large racecourse where one might see thoroughbreds imported from England. As it grew richer and bigger, however, it became tamer. Gambling and Sunday labour were prohibited, and in time the mine workers became unionized. Billy Barker presided as leading attraction and Father Christmas only until selling out his stake in 1864 and falling to an ever-worsening series of jobs and situations; he died of cancer in Victoria in 1894. A fire in 1868, the sort that so often befell remote wooden towns without waterworks, destroyed much of Barkerville, but the community lingered on because of the Cariboo Gold Quartz Mine Co., a large dredging firm that had consolidated many smaller interests. It continued to operate a dredge there as late as 1966, when a restored Barkerville was already known as a tourist draw – a far less overrestored town than Dawson in the Yukon.

The importance of the Cariboo gold rush in shaping Canada, by encouraging the idea of a transcontinental railway and the concurrent dream of Confederation, is easy to underestimate. So is the role of the Cariboo in the great multi-ring circus of the gold crusade, an importance often most easily represented by single individuals whose experiences were typical, whatever their personal degree of success. There was, for example, Jacob W. Davis (born 1831 as Jacob Youphees at Riga, on the Baltic). He was a tailor who joined the California gold rush via Panama in 1856 and two years later followed the rainbow to the valley of the Fraser, where he began to make trousers of white duck, and later of denim, using rivets to reinforce the seams. Later, back in California after British Columbia faded, he shared the patent with the more famous Levi Strauss. And then there was Richard Carr.

Carr was an Englishman, born in 1818, the son of a tradesman. In 1837 he emigrated to the New World, which he saw in a series of often unrelated jobs that took him from Lake Michigan to Lake Nicaragua. In fact, he was in the fortunate position of already being in Central America when he heard news of the strike at Sutter's Mill. He arrived in San Francisco in January 1849, working as a daguerreotyper (everyone, it seemed, wanted a likeness to send back home) and then trying his hand at the diggings. In a year's prospecting he had saved only $3,000, not much by gold rush standards, but he invested it in general stores in San Jose and other localities in the gold region. By 1854 he was worth $25,000, and in 1861 he returned to England and retired to a manor in Devonshire, where he grew bored.

In 1863, when the Cariboo mines were at their pinnacle, Richard Carr took ship from San Francisco to Esquimalt on Vancouver Island, from where he alternately rode and walked to Fort Victoria, whose population stood at two thousand Coast Salish and three thousand assorted whites. Having made one fortune in the previous rush, he proceeded to make a second in another, this time by setting up as a wholesaler and commission merchant. But by 1865 the gold resources, or such as could be reached by even the best-capitalized syndicates, had been badly depleted, and the civil war in the United States had played havoc with the markets and the labour force. The Cariboo rush was fading fast, and Victoria lost its status as a free port, sealing the fate of such businessmen as Carr. He sought solace in religion and lived until 1888. He never knew that his daughter Emily Carr, only seventeen when he died, would become one of Canada's greatest painters, the one who took inspiration from and revived interest in the culture of the West Coast Native peoples to whom the gold had belonged originally.[2]

Whenever gold was gathered in a few extremely rich concentrations, gold rushes were likely to be highly dramatic and short-lived, because of the sheer intensity of the occurrence. In other places, where the gold was more evenly distributed, the result was more likely to be a series of smaller rushes. These serial rushes were frenetic because of their reduced scale, each siphoning energy from the last and quickening everyone's hopes. Such was the pattern in New Zealand, where the effects of gold differed substantially from the Australian experience, even though many prospectors crossed the Tasman Sea travelling east or west from rush to rush.

The gold rushes in New Zealand did not foster the same sense of exploration as those in Australia, for the land was already settled. Gold never displaced sheep ranching as an economic cornerstone, nor could it approach the Maori Wars as an event of nation-shaping importance. But the sociology of it is fascinating because New Zealand was determined not to be bettered by Australia and used its gold to lure development. New Zealand profited by the examples of Australia and California while maintaining peace and order in a way that Australia had not done and California had not even thought of.

In 1852, with events in Victoria at their peak, a committee of civic leaders in Auckland offered a purse for the discovery of a payable gold-field in the North Island. Later and in the same spirit, the provincial government of Otago in the South Island dangled £500. There were flurries of activity and claimants for both prizes, but few people bestirred themselves until 1857, when hundreds of prospectors began working claims in the Aorere River. The organic character that would mark the New Zealand gold rushes was apparent even then. There are accounts of how the miners relaxed with music at the close of day and shared a common evening meal. A more effective incentive than any cash prize was the Goldfields Act passed by the New Zealand legislature in 1858. Taking the relevant Victoria statutes as its basis, it imposed an export tax on gold but established a fair system of miners' rights or licences. It was thus in place when an Australian, Gabriel Read (1824–94), made the first large and significant discovery in 1861, a time when many Australian prospectors were having to flee or accept their fate as wage slaves of the big mining concerns.

Read, who was from Hobart and had had experience in the California diggings as well, struck it lucky at Tuapeka in southern Otago, though at first there was widespread belief that the news of his find had been concocted by the shipping lines to benefit the passenger trade. But the evidence of the strike was far more tangible than any suggestion of conspiracy could be, and in July a couple of thousand miners crossed from Victoria in a heightened state of anticipation if not frenzy. Otago at that time had a population of thirteen thousand; by October the figure had doubled. At the site of Read's discovery a town called Gabriel's Gully grew up, with places of entertainment that included sly grog-shops, even though the sale of liquor was opposed by the ad hoc miners' committee. It was with the committee that authority rested until the government could appoint officials, including a police commissioner who had had experience in Victoria.

By statute, claims were limited to thirty feet in order to preserve the appeal to independent prospectors, whom the government hoped would be converted to settlers eventually (though few were). The winter of 1862 was particularly harsh on the estimated seven thousand people in the goldfields, with heavy snows and with wood so scarce that what little could be scavenged had to be used for cooking rather than burned for warmth; even the stick used to stir the billycan was carefully saved and used again and again. Many people went home to Victoria. For those who stayed, the government continued to exert itself in being attentive to their demands, which included more townships (a concern of shopkeepers, who feared they might not have clear title to their plots), a coal supply, better roads, a hospital, and a government gold escort. Some of the public-works projects used labour drawn from the ranks of unsuccessful miners, but it was difficult to keep prospectors from rushing off somewhere else.

In August 1862 the Cobb & Co. coach arriving from Dunedin on the coast brought news of gold along the Molyneau River at the base of the Dunstan Mountains. Two former forty-niners, Horatio Hartley (died 1903) and Christopher Reilly, had hit upon a rich area and had to be bribed with a £2,000 reward into revealing its exact location. Once they had done so, the rush was on. One who came was James William Robertson (1824–76), a native of Saint John, New Brunswick. He had mined at Ballarat and Bendigo and proceeded to make his fortune in Otago by supplying the miners with flour and lumber from mills he owned.[3]

Once again, relations between the miners and the authorities were particularly harmonious, with the police constables escorting the merchants' supply trains and forcing the owners to limit the extortion in their system of pricing. Other discoveries flared up all around. A miner named Edward Fox (died 1875) left the Dunstan for a more remote area, but he was followed and his secret discovered. He assumed control of the situation, however, by calling himself the commissioner and doling out sixty-foot claims. By the end of the year, there were three canvas towns in the Dunstan, and as such places go they appear to have been full of good fellowship, with sporting events and a communal feast at Christmas. Again, the government made attempts to persuade miners to stay. Provided they had six months' residence, the holders of miners' rights were permitted to vote, for instance. But if in Victoria the diggers at first had less democracy than they demanded, the ones who came to

New Zealand received more than they cared to exercise, perhaps because the serial nature of the small gold rushes left them feeling they had no permanent stake in the country. The conspicuous exception was Julius Vogel (1835–99), a gold rusher turned editor, who led the never successful movement to separate the two islands but later headed the government of New Zealand anyway. In 1862 Otago had a population of approximately sixty thousand, of whom as many as twenty thousand had come for the gold. Of that number, perhaps sixteen thousand departed over the next two or three years. One reason was the trend towards amalgamation and capital, forces that were able to lobby against the interests of the small-scale miners. The bigger factor, however, was simply the better or at least fresher opportunity offered elsewhere, as in 1865 when four thousand people joined a rush to Hokitika on the island's west coast.

The loss by one provincial government was another's gain. The authorities of Canterbury had been posting rewards and sponsoring surveying parties ever since vast new tracts had been purchased from the Maori in 1860. The efforts did not return a dividend until mid-1864, when specimens of gold taken from a tributary of the Teremakau River were exhibited at Christchurch with predictable results. Thousands hurried in from Otago. Others arrived with packhorses from Canterbury over a difficult mountain pass. Still others came by sea from Wakamarina, braving shipwreck in the dangerous surf at the end of the journey or on the treacherous sandbar that blocked the mouth of the Hokitika River. There the town of Hokitika sprang up. It had many of the characteristics of gold rush towns around the world: the all-pervasive sense of impermanence, the architectural rawness, and the ludicrous displays of luxury that seemed so out of place in a spot remote from genuine civilization. It even offered the usual jumble of races and languages, for it attracted some Germans, Frenchmen, Greeks, and Italians, and a great many Scandinavians. Only the violence was lacking, a fact attributable to the absence of Americans and a prohibition on weapons. The few incidents of bushranging were attributed to the Australian example.

The community was not especially cohesive, however, for as in California and, later, in the Klondike, the miners were strung out over a considerable distance, working the various creeks alone or in twos and threes. This led to travel farther up the streams and so to more discoveries farther afield, the first being about nine miles north of Hokitika. In a scene that prefigured later events at Nome in Alaska, thousands of miners began working the beach at Okarito, fifty-five miles south of Hokitika, when gold was discovered mixed with black sand right above the tide line; it was so easily gathered that the camp was worked out in six months. The last hurrah of the New Zealand gold rush, so far as small prospectors were concerned, took place in May 1867 when a black man, whose name is now lost, started a rush eight thousand strong to the Buller Range, while others headed to Auckland province on the North Island.

As the rush wound down, many New Zealanders joined the hordes of returning Australians. Their destination was Queensland, which experienced its own series of brush-fire gold rushes, starting in the late 1860s. Ranching and agriculture were already important in Queensland (whose residents the newcomers called bananamen, though sugar cane was the more important commodity). So, however great the social disruption, gold did not affect the economy as forcefully as it had done in Victoria. Like New Zealand, Queensland adopted a variation of the Victoria system, exacting an import duty in return for providing such services as a gold escort and a system of miners' rights.

In Queensland, claims were either fifty square feet or fifty by one hundred feet and were leased for twenty-one years. But the nature of the geology there – less free gold – meant that syndicates and companies were more influential from the start, so the whole colony was less attractive to diggers than New South Wales or Victoria had been. Because the competition for the gold was so intense, there was pressure to put racial minorities at a disadvantage, and Queensland, though it lacked California's propensity for violence, did no better than California in its treatment of the Chinese. First, they were simply chased from the camps; then they were charged twice as much for a miner's right as white men and more than three times as much for a business licence, though the Crown finally stepped in. Mindful as everyone was of the events at Eureka Stockade, the authorities in Queensland were careful not to excite the passions of the white miners. They gave male miners the right to vote after six months' residency. But minorities were no such threat and the authorities saw no need to contradict the popular fear and hatred of them. In 1877 a ten-pound tax was imposed on Chinese, and some ships were turned back altogether. Then Chinese as well as black miners were excluded from the goldfields for three years from the date of first discovery. It was thought that the alluvial ground was too precious, too easily exhausted, to be squandered on other races. From the diggers' point of view, it was bad enough that the big companies were beginning to cast their corporate shadow over the goldfields.

4

Silver into Gold

All the new gold being mined was pumped into the world's money supply and provided a great boost to speculation. Credit was easy even
though the international economy was stretched tightly. The first tear
came in the United States. In August 1857 an Ohio insurance company
failed, touching off the most serious financial panic in twenty years.
Banks closed, businesses faltered, trade was crippled; ordinary citizens
were brought up short in the realization of just how inextricably their
lives were bound to the large abstraction called the economy. The California gold rush had wound down by then, but the others that followed
it carried a new edge of desperation. One saw this in the scramble to the
Fraser River Valley and the Cariboo. One saw it in all the other rushes,
large and small, that lit up the western half of the continent like brush
fires for the next half-century. Imagining North America as the human
body, one might say that the *adrenal medulla* responded to the slightest
provocation with full production.

Having gone as far west as possible and then as far north as their
imaginations permitted for the moment, the Americans began to double
back across their own country, frantically seeking wealth almost as
though they could will it into existence, proving again that the phenomenon depended not so much on the amount of gold found as on the state
of people's minds. But in the Cariboo, too, the reality had proved
unequal to the make-believe. By the mid-1860s it was clear that the
rushes had taken on a different coloration through the emergence of
what one historian has described as 'the class of habitual or professional
prospectors, recruited from all parts of the world ... permanently
afflicted with mining fever, always hopeful though seldom successful ...
to be found in every rush to each new section where gold was said to

have been found.'[1] The diggings resembled nothing so much as the
pirate settlements once found in the Caribbean and the Indian Ocean,
outposts of grass-roots anarchy, composed of all nations and races, living under no code save the one they concocted to maximize trade and
ensure freedom from convention.

The financial panic was solidifying into a recession when in the spring
of 1858 William Green Russell led a group of prospectors into the
remote section of what is now Colorado, along Cherry Creek. The group
consisted of individuals from Russell's own state, Georgia, and others
who had joined them at the Missouri River, as well as some members of
the Cherokee Nation. They knew where to look because Native people
had been coming down out of the Rockies for years with stories of gold
there, but the inherent racism of the time had caused most who heard
these stories to dismiss them. Russell's party and another one that had
gone in from Lawrence, Kansas, both made small discoveries easily
enough and then disbanded. In going their separate ways, the prospectors only helped to spread the reports more quickly, and the stories
increased in size as they gained velocity. Before autumn, the Cherry
Creek area was full of gold seekers, who immediately founded two
towns, Denver and Auraria, which later merged under the name of the
former. The best-known topographical feature of this wild and unpopulated region of high fir-covered slopes was Pike's Peak. The phrase
'Pike's Peak or bust' entered the common language the following spring
when the Colorado gold rush got under way in earnest.

The most useful witness to the Colorado rush, the equivalent of
Bayard Taylor in terms of completeness and veracity, was Henry Villard
(1835–1900), a Bavarian who had changed his name from Ferdinand
Heinrich Gustav Hilgard when he emigrated in 1853 in order to avoid
military service. He became a reporter for various New York newspapers and later bought control of the *New York Post*, having by then made
his name as an important railway promoter, general capitalist, and the
financial backer of Thomas Edison. People in the East found the thought
of a Colorado rush especially appealing because, compared with California, the distance was shorter by one-third and involved neither
desert nor the worst in mountains (Cherry Creek and the other sites
being well on the eastern side of the Continental Divide). Yet more
hardships awaited them than they supposed, thanks in part to the way
Missouri River outfitters, with a vested interest in promoting a rush,
sold them the wrong sort of equipment and supplies.

'It is doubtful whether such scenes of human misery ... were witnessed even at the height of the California excitement,' wrote Villard in his book *The Past and Present of the Pike's Peak Gold Region*. Indeed, once they arrived, many of the wretches promptly packed up and returned home. There is no knowing what level of regret they felt when in May 1859 a Cherry Creek miner turned up in Denver with evidence of pay dirt. Villard heard the man's story and witnessed the scene it touched off: 'Traders locked up their stores; bar-keepers disappeared with their bottles of whiskey, the few mechanics that were busy building houses, abandoned their work, the county judge and sheriff, lawyers and doctors, and even the editor of the *Rocky Mountain News*, joined the general rush.' By the middle of June, there were ten thousand people taking part, and one rather large town, at first called Mountain City but soon renamed Central City, appeared in what seemed like an instant. By summer, however, a great number of the miners had gone home. There were profitable placer sites, particularly at Jackson Diggings near South Clear Creek and later on the other side of the divide at South Park and Blue River diggings, but for the most part Colorado meant deep shafts and companies to sink them. The same end result – so dispiriting to many – was also true of the other such event taking place at that time: the Comstock silver rush in Nevada.

The parallels are obvious and the differences illuminating. A number of forty-niners had crossed the California boundary to the Washoe Lake region of Nevada, convinced that there was gold to be found. The areas around what became Carson City and Gold Hill looked especially promising. But no one had given much thought to silver, even though it exists in similar geological conditions and is usually found in combination with gold or other minerals. To deliberately search for silver in an age of gold rushes demands more than a contrary disposition or shrewd deviation from prevailing trends; it requires a more realistic set of mind, something that can develop only after the mass optimism has begun to turn sour, making desperation more acute. In 1856-7 four partners who had become so convinced looked for silver deposits near Gold Hill and found them, but quickly came to grief. One of the four was killed in an accident, a second dropped out owing to broken health, a third died of exposure in the mountains, and the last was killed by bandits. Accordingly, there was no stampede at first, only a steady egress of miners still searching for gold.

Horace Greeley, the New York editor, reformer, and crank, who in 1855 had urged young men to go west, was in Placerville, California, in 1859, urging them to backtrack a bit to the east. Across the desert, he opined, there 'may be a land of vast mineral wealth.' And so there proved to be when, on 12 June 1859, Peter O'Reilly and Pat McLaughlin, Irish immigrants working about eighteen miles south of where today the casino lights of Reno cast an artificial glow on the horizon, found an enormous ledge of the bluish ore that previous prospectors had viewed with disdain. O'Reilly and McLaughlin still thought of the discovery in terms of gold until an ex-farmer with more perspicacity took a sample to be assayed. The report showed that a ton of the ore would produce $1,595 in gold but an astounding $4,791 in silver. And there were tons of the stuff almost beyond counting.

By that time, another prospector had announced that he had been working the same property all along and deserved a share. He was Henry T.P. (for Thomas Paige) Comstock (1820–70), a native of Trenton, Canada West. He had indeed been prospecting in Nevada since 1856 but was a mule skinner by profession and has come down in history as something of a charlatan, 'a sanctimonious gaffer' in the telling phrase of one historian. So began an exhaustive legal fight between Comstock and a number of others, including Richard Maurice Bucke (1837–1901), later to be famous as a Canadian psychologist, as the author of *Cosmic Consciousness*, and as Walt Whitman's friend, biographer, and literary executor. Thus, too, the beginning of what came to be called, universally but unfairly, the Comstock Lode, from which about $145 million in silver was taken over the next decade. In the melodramatic pattern of such events, however, the discoverers and the discoverer presumptive did not stay rich for long. The spot the three men owned became the Ophir Mine, which in time yielded $50 million in silver. O'Reilly sold his share for $40,000, or twice what Billy Barker received in similar circumstances, and like Barker, he squandered it; he died in an asylum. McLaughlin got only $3,500, sank to working as a cook on a ranch, and was buried in a potter's field. Comstock, who later laid claim to what became C Street in Virginia City, got $11,000. He stayed in Nevada, a presence conspicuous by his self-aggrandizing, until 1862. He committed suicide with a revolver.

Like California before it, Nevada by its name alone suggested bravado, freshness, an attractive and creative confusion. Even the language spoken was put together hastily to suit the occasion. As late as 1888, Walt Whitman, in a magazine article later included in *November Boughs*, cackled with delight in quoting a typical piece of prose from a Nevada newspaper: 'The toughest set of roosters that ever shook the dust off any town left Reno yesterday for the new mining district of Cornucopia.

They came here from Virginia. Among the crowd were four New York cock-fighters, two Chicago murderers, three Baltimore bruisers, one Philadelphia prize fighter, four San Francisco hoodlums, three Virginia beats, two Union Pacific toughs, and two check guerrillas.'

The three Nevada towns associated with the Comstock rush are Gold Hill, Silver City, and Virginia City. They arose with all the spontaneity and lack of planning associated with mineral rushes. A traveller in 1865 might have been speaking for such mining towns generally when he found in a satellite community, Austin, Nevada, that 'houses are built anywhere and everywhere, and the streets are then made to reach them.' Yet this unprepossessing place, he added, 'has the best French restaurant I have met since New York [and] the boot-blacks and barbers and baths are luxurious and aristocratic to the continental degree.' Virginia City soon became the biggest city in the entire West after San Francisco, and only San Francisco could rival its queer juxtaposition of squalor and splendour.

Virginia City was named by Jack Finney, a prospector who was also known as Old Virginny. He christened it while drunk as usual, though whether in honour of himself or his native state remains unclear. Finney owned a valuable claim there but sold out, says the legend, for a mule and a quart of whisky, and at that he was luckier than many of the four thousand other miners who had come to Virginia City by the close of 1859. A visitor the following spring described the city this way:

Frame shanties pitched together as if by accident; tents of canvas, of blankets, of brush, of potato-sacks, and old shirts, with empty whisky barrels for chimneys; smoky hovels of mud and stone; coyote holes in the mountain-side forcibly seized and held by men; pits and shafts with smoke issuing from every crevice; piles of goods and rubbish on craggy points, in the hollows, on the rocks, in the mud, in the snow, everywhere, scattered broadcast in pell-mell confusion, as if the clouds had suddenly burst overhead and rained down the dregs of all the flimsy, rickety, filthy hovels and rubbish of merchandise that had ever undergone the process of evaporation from the earth since the days of Noah.

The tone of extreme dejection was not misplaced. As in Colorado, both then and in the future, miners learned that Nevada required stamp mills to crush the ore, as well as other costly and complicated installations. Hence the many instances of single miners selling their claims to others who could put them together with claims on adjoining property to make them good credit risks. It was an environment for enterprising middlemen, all of them hoping to assemble larger or more profitable pieces of the total puzzle and so become overlords. The wealth thus created transformed Virginia City so quickly that the traveller quoted above would have seen imposing streets of brick and stone buildings had he returned a couple of years later. He might also have seen evidence of the locals' desire to buy respectability in the East. A miner named Big Jim Fair went so far as to build a mansion for his daughter in Newport, Rhode Island, the enclave of Cornelius Vanderbilt and others of that type.[2]

The wealth may have been concentrated in fewer hands than had been the case in California, but the general demographics and sociology were the same. The overall level of sophistication is surprising considering that Virginia City was, and remained for such a long time, an almost wholly male society. In its first days, there was but one woman in the community, Eilley Orum, a Scot who had outlived two Mormon husbands and made a living by letting rooms and taking in laundry. She next married Sandy Bowers, a miner whose ten feet (claims were now measured in linear feet) made him very rich, and she spent most of her remaining years traipsing through Europe, being snubbed by royals and aristocrats for her ghastly manners and lack of poise. The second woman in Virginia City was Julia Bulette (1826?–67), a much-loved prostitute from New Orleans, though Liverpudlian by birth. The other citizens once had to defend her, and the town, from an aboriginal raiding party, and she reciprocated by nursing them during an influenza epidemic. She presided over the tenderloin district, Sporting Row, but came to a tragic end. The public hanging of the man who murdered her during a jewel robbery was Virginia City's first great social event. The Pullman car named in her honour by the Virginia & Truckee Railroad (completed in 1869, the same year as the transcontinental line) was later bought by Paramount Pictures for use in westerns.

Probably at no time in the formative years did the female population of Virginia City or similar mining towns exceed 10 per cent, a situation calculated to intensify what the Victorian miners thought of as chivalry. The actress Adah Isaacs Menken (1835?–68) was a favourite of the period whose most famous role was in *Mazeppa*, at the conclusion of which she exited stage left wearing practically nothing and bound to a black horse. Such was the admiration of a Virginia City audience that it made her a gift of a silver bar worth $2,000. Folklore has it that the bar was so heavy that Wells Fargo & Co. had to reinforce the railcar on which it was shipped to Reno.

Generosity on that scale now seems an inversion of the attitude that produced the violent behaviour of people such as Fighting Sam Browne, who is said to have carved the heart from a man who disparaged his performance in a bar-room brawl. Yet the violence is infinitesimal and certainly less the product of derangement than that to which later generations have become accustomed. There was something wonderfully innocent about such places, as attested to by the high degree of what the age called vice. In Virginia City today, visitors are shown the suicide table at the Delta Saloon, so named during the boom years when, it is said, a faro player broke the bank and the three proprietors took their own lives in response. Incidentally, one can still read the *Territorial Enterprise*, where Samuel Clemens worked after failing to strike it rich in silver and where he first took the name Mark Twain. And one can still see what in the loose usage of the day was called the opera house, where, in this case, Sarah Bernhardt, Lillie Langtry, and Enrico Caruso performed.

Politically and economically, the Comstock Lode had enormous significance. It provided the federal government with a new source of capital with which to finance the war against the southern states. Silver grew more important when it performed well during the gold panic that began on Black Friday, 24 September 1869. As it had done in San Francisco, the government waited until the rush gave way to stable production before establishing a mint at the site. The reserves of Carson City silver dollars were so large that Washington could put many hundreds of thousands of pieces on sale as late as 1980. Whereas profits from the California rush found their way to Britain and France as well as the eastern states, the Comstock Lode was of benefit mostly to the West and westerners. San Francisco, for example, matured and was made permanent by this second great infusion of wealth.

It was after making a fortune from the Gould & Curry mine that George Hearst (1820–91) bought himself first the San Francisco *Examiner* (laying the groundwork for his son William Randolph Hearst) and then a seat in the U.S. Senate. Another noted Comstock Lode fortune was that of John W. Mackay (1831–1902), an Irish-born miner who arrived in California in 1851 and later, hearing of the Comstock Lode, walked the 225 miles from San Francisco and got work wielding an Ames shovel, a brand name so common that it was almost a generic term. But typically for that time and place, his fortune came not as a miner but as a deal maker. In 1863 he heard of three men who wanted to develop an important mine property they owned but who lacked the consent of a fourth

party, without whose share of the claim their own pa.
Learning that the missing man was fighting with th
armies in the South, Mackay tracked him down and, unde.
the man's share for $500. He later became part owner of a .
the Big Bonanza (*bonanza* being the Spanish term whose
borrasca, was seldom if ever used, despite the many possibl.
cations). Mackay married the widow of a gold rush physician
Downieville, California. His fortune was so large and durable
adventurers had to be discouraged, and in the 1920s Mackay's son dis.
herited his own daughter when she married Irving Berlin, though th
songwriter was already worth millions himself.

Another telling connection to California was Adolph Sutro (1830–98),
a Prussian engineer (or cigar maker when down on his luck) who turned
up in California in 1850 and in Nevada ten years later. By 1869 many of
the big mines were closing because of flooding. Sutro conceived the idea
of a tunnel through Mount Davidson, on whose slope Virginia City
reposes, that would drain the underground streams and allow renewed
access to the ore. Local capitalists first rebuffed him, then intrigued
against him, and finally accepted him. It took a decade to complete the
project, which continued in use well into the 1940s. Sutro sold his interest partway through and invested in real estate in San Francisco, where
he also served as mayor.

The tunnel's importance is that it illustrates how even the simple
extraction capitalists of the first generation were displaced by changing
technology as surely as they themselves had displaced the lone prospector with his shovel and dish. Nineteenth-century mining methods were
inefficient and sacrificed everything to volume. Thus, many mining
towns were honeycombed with shafts and tunnels (seven hundred
miles of them in Virginia City), while the townscapes were steadily displaced and subsumed in mountains of tailings, themselves a great
source of wealth to the first people who learned how to sift them effi-
ciently. But with each new fortune created from an existing one, there
was less and less room for individuals to start from scratch. The hordes
of argonauts who had gone to California to become independently
wealthy, or at least to remain independently poor, were now employees
working for wages. Every mining town that had its obligatory opera
house soon had its union hall as well. The desperation of individuals
acting en masse was now the very different and infinitely uglier desperation of opposing economic forces.

Bernard Baruch, the Wall Street financier and self-styled adviser to

American presidents, made part of his fortune in a Nevada gold mine at the turn of the century, backing George Wingfield, a mining man who had bought up the claims of individual prospectors and consolidated them with $1 million from Baruch and others. Baruch's favourite story, recounted in all the biographies and memoirs of him, concerns the way Wingfield and his armed mine guards would greet the shift each evening, forcing each worker to strip and then jump over a high bar, so that any gold he had concealed in his anus would fall to the floor. The practice of stealing small amounts of gold was called high-grading. Wingfield's reaction to it brought the attention of the Industrial Workers of the World, but their attempt to organize the miners was violently suppressed. There is reason to believe that Wingfield framed two Wobblies on murder charges in 1907, causing the unrest that led the federal government to send in troops, who crushed the union. His two victims received posthumous pardons in 1987.[3] The spirit of confrontation was to be played out again and again throughout the mining districts.

As gold rushes turned corporate and forced out the free spirits, other potential opportunities opened to them; and so across the western states and western provinces a series of smaller gold rushes – or, really, one long intermittent rush – took place. On the heels of Nevada came word of strikes on the Columbia River in Washington, igniting a dream in which there might yet be room for lone placer miners. Next came Idaho, where, in similar fashion, the relatively small discoveries were enough to attract a certain type of person for years to come. In the 1890s, long before he created Tarzan, Edgar Rice Burroughs was a gold prospector in Idaho, but he reported that he gave it up because he consistently lost at gambling what he made at mining. The tiniest and most respectable modern communities can conceal gold rush origins. In 1869 a thousand miners rushed to Julian, sixty miles east of San Diego in California, at the first hint of placer gold; the town is now almost synonymous with apple orchards.[4] Montana also had its gold boom, though with prosperity came respectability of sorts, and the community of Crab Town was renamed Helena. So too the Dakotas. Even Ambrose Bierce, the vitriolic San Francisco journalist who had all but built a career on ridiculing the old forty-niners (and later mocked those foolish enough to head for the Klondike) was caught up in the excitement and hurried to the Dakotas to make his fortune.

It was the influx of prospectors into the Black Hills that raised the ire of the Sioux and other Native peoples, and proved so fatal to Colonel George Custer, who had been sent in to protect the miners. His defeat took place in 1876, the same year as a notable murder in Deadwood, the local mining capital. Drawn by the lure of easy money that gold towns suggested, a sometime miner, sometime gambler, and sometime policeman named James Butler Hickok, popularly called Wild Bill, turned up in Deadwood, which was then enjoying its turn at all the outlandish attractions of earlier gold rush towns. Hickok became quite a figure, dressing foppishly to suit the mood, even allowing a play about his exploits to be performed. But he lost sight of the gritty reality that gold rush glamour was beginning to subvert and was shot in the back as he sat playing poker. His murder did much to spread the renown of Deadwood, home of the Deadwood stagecoach, which Buffalo Bill Cody, who also worked as a prospector when conditions suited, immortalized in his Wild West shows.

Far to the south, late in the century, silver-mining booms overtook New Mexico and the Arizona Territory, giving Tombstone the same resonance as Deadwood. But these were all holding actions against the revival of the gold rush in Colorado, which came about as the nineteenth century – and the type of independent behaviour it encouraged and tolerated – was reaching its conclusion.

As there was some gold in Nevada, so was there some silver in Colorado, but the pattern was always the same – a discovery, followed by ballyhoo and the overnight erection of a town to which thousands hastened, only to find that a few very early arrivals had become very rich and that the town was virtually a company town. In 1877, for example, silver strikes were made at Leadville, Colorado, with predictable results. After a brief struggle for supremacy, Horace Tabor (1830–99), owner of the Matchless Mine, emerged as Leadville's richest and most powerful figure and also, perhaps somewhat redundantly, as its mayor. Leadville briefly held the title of the richest and most wicked city in the land; but such was the level of censorious moralism that not even Tabor could escape being ostracized when in 1883 he divorced his wife to marry the widow of a Central City miner. The couple scandalized the entire region and were forced to live behind drawn draperies. The wealth evaporated with the free silver movement of the 1890s. Elizabeth Doe Tabor survived her husband and lived in a shack near the entrance to the Matchless Mine until she froze to death in 1935. An opera has been written about their life together.

Oscar Wilde touched at Leadville on his 1882 lecture tour and found the miners there 'charming.' After seeing the way he drank and smoked, they responded by calling him 'a bully boy with no glass eye,' a statement Wilde called 'artless and spontaneous praise which touched me more than the pompous panegyrics of literary critics ever could or did.' Wilde also recalled, 'I read them passages from the autobiography of Benvenuto Cellini and they seemed much delighted. I was reproved by my hearers for not having brought him with me. I explained that he had been dead for some time, which elicited the inquiry, "Who shot him?"'[5]

The largest Colorado mineral rush, and the one with the most lasting effect, came later and centred on the community of Cripple Creek. In 1891 a cowboy named Bob Womack, who had been prospecting part-time for fifteen years, found a rich vein of gold on land his father had homesteaded in 1876; he found it at a place called Poverty Gulch. Gold was $20.67 an ounce at the time, and until the vein became uneconomical for the technology available to exploit it, some 500,000 ounces were taken out. True to type, Womack sold out for $500 and later became part of the workforce of common miners; he died penniless in 1909. The purchasers were two Denver real estate operators and cattle barons, Horace Bennett and Julias Myers, who promptly platted an eighty-acre townsite for a community they called Cripple Creek, which was incorporated in 1892. Before the year was out, the town had seventy-two lawyers, forty stockbrokers, and thirty-nine estate agents, as well as its own stock exchange, and restaurants and chemist shops that ran all night. In the parlour houses and the less expensive cribs along Myers Avenue there were perhaps as many nationalities of prostitute as there were of miners working underground. It is fitting that the song 'There'll Be a Hot Time in the Old Town Tonight' was composed in Cripple Creek.

Two other towns – Victor, where the journalist Lowell Thomas grew up and first worked as a reporter, and Gold Hill, a suburb of Victor – completed the Cripple Creek region, which reached a peak population of about thirty-five thousand at the turn of the century, following several disastrous fires and the usual assortment of booms and recessions. Only a few streets remain of the towns today, because the communities choked in their own tailings and because, in company towns, when the minerals were played out, the owners razed buildings in order to reduce property taxes and limit insurance liability.

The most important of the estimated thirty millionaires resident in Cripple Creek was Winfield Scott Stratton, a Colorado Springs carpenter who, like Womack, struck pay dirt after fifteen years' searching. Stratton called his mine the Independence because he had made the discovery on 4 July 1891. It was one of the famous mines of the nation, like the Gold

Coin Mine in the heart of Victor, which was discovered during the course of excavating for a new hotel. Stratton took out $1 million in gold before selling out to British interests for $11 million. The property ultimately generated $28 million in gold. As the rush lengthened, many new nationalities were to be seen, including Slavs and eastern Europeans. Cornishmen, who had been a fixture in the earliest copper rushes in the United States and Australia before there were gold rushes in either place, were present in considerable numbers; as at Ballarat or any other established mining centre, they were called Cousin Jacks. The name was also given to one of their tools, a small wheelless barrow used in low crevices underground.

To promote a quality workforce and also to forestall high-grading, Stratton paid a wage of three dollars a day. There were even improvements in mine safety, though they came slowly. One of the simplest and best was the safety cage in mine lifts, which had steel hooks that dug into the rock and earth if the cage began to slip. It was invented in Australia by Joseph Hillerman, who had redesigned a traditional Cornish mining tool into something universally known as the Ballarat pick. But high wages and more attention to safety could not prevent organized labour from entering the mines. As the years passed, labour wars became endemic in Colorado, especially in the copper mines and coal mines. The starkest clash was the Ludlow Massacre of 1914, when thugs employed by the Rockefeller coal interests gunned down children, women, and unarmed men. It is significant, though, that Woody Guthrie who, many years later, wrote a famous protest song about the incident, was himself a victim of gold rush fever in his youth, running off to Texas with his family when they heard news of a strike there.

Colorado gold mining had entered a long period of decline, despite a small strike in 1904. Many mining companies merged; other shafts were closed because flooding began to occur even at three thousand feet. The lessons had already been learned by miners and others in two far more important rushes at opposite ends of the world – the rush in southern Africa, where the individual lost out to the company and the state, and that in the Yukon, where the lone miner was often the winner and some of the wonderful promise of 1849 was at last recaptured.

5

Titans in South Africa

In 1850 South Africa was so isolated, so underdeveloped, and so thinly populated that only one of its citizens can be shown to have taken part in the California gold rush. Nevertheless, that solitary figure, one Pieter Jacob Marais, would be enough to carry on the crusade and bring gold fever to another continent. But the gold rush which he sparked in South Africa, and which was by any standard the world's largest, was less the work of a single instigator or publicist than past rushes had been or later ones would prove to be. What's more, Marais (1827–65) is atypical in that he was the lone prospector of the stereotype, with limited resources and great resourcefulness, effecting his small pact with the government but determined to make his fortune from scratch. He could not be more unlike the figures who dominate the story of the South African rush, men who for the most part had already made their fortunes in the diamond fields and were thus able to bring to deep gold mining the vast capital demanded by South African geology and power politics. Other gold rushes had held out for the ordinary person the promise of financial independence or at least well-earned self-respect. South Africa's gold rush was a ruthless game for millionaires, who played it from low motives and for high stakes. In no other country was the social cost of gold higher, and nowhere else are the scars still so visible.

Because southern Africa had known European settlement much longer than California, Australia, New Zealand, or British Columbia, the stray rumours of gold discoveries had a longer history there. They date from the first generation of Dutch settlers who in 1652 had founded Cape Town (which was taken over by the British in 1795, returned in 1802, captured again in 1806, and formally ceded to Britain by the Netherlands in 1814). The Boers themselves had heard stories of Arab traders who, perhaps for centuries, had been acquiring small amounts of gold from tribes in the interior, and there was ample evidence of primitive precolonial mining. But such tales and ruins stirred no money lust in the sternly sober and Calvinistic Boer farmers, who in any case had more pressing concerns. Until the middle of the nineteenth century, there were periodic wars with the wide range of tribes who spoke the various Bantu dialects; these conflicts were known as the Kafir (or Kaffir) wars, from an Arabic word for infidel. The distrust which the British and Boers felt for one another grew particularly strong in the 1830s, the era of the Boers' Great Trek out of British territory and across the Vaal River to an area thenceforth known as the Transvaal. Such was the informal name of the South African Republic, the homeland the Boers established for themselves by defeating the Zulus (though they were to endure a further period of British domination, between 1877 and 1881).

As had been the case in the United States and then in Australia, a small copper rush gave a taste of what the later influx of miners would be like. This preview took place in the Cape Colony in the 1850s, a time when there might have been gold rushes near by if they had been allowed to develop. In 1852 a Welshman named John Henry Davis and his son Henry John, who had hunted diamonds and gold in Brazil, went searching the watershed of the Transvaal Plateau, an area called the Witwatersrand. They made some discoveries of gold, but the news was hushed up, and there the matter rested until Marais hove into view.

Marais had departed for California only one week after the news of Sutter's Mill reached his native Cape Town. It was a fearsome journey, via England, but the Marais who appears in photographs – a short, solidly built, no-nonsense sort of fellow with a large spade-shaped beard – was apparently made of sturdy stuff. He first found work as a clerk in a San Francisco shop before panning for gold along the Yuba River with moderately good luck. At length, he was successful enough to start his own business, but it burned, along with much of San Francisco, in October 1850. Undeterred, to judge from the matter-of-fact tone of his diary (though complaining of bowel trouble as usual), Marais left for Australia at the close of 1851.

In Bendigo, too, he experienced a small degree of success, and one day when selling his gold in Melbourne he chanced to hear some men speaking with South African accents. A shipload of prospectors had just arrived from the Cape, one of them bearing a letter for Marais, possibly telling of his father's death. Whether for that reason or because he was nostalgic or because he wished to try out his now considerable prospecting expertise in his native colony, Marais decided to take ship for Cape Town. He found employment in the copper fields and then, in September 1853, crossed into the Transvaal to prospect there. On 7 October he found specks of gold on the Crocodile River (recording the event in his diary in uncharacteristic italics).

This discovery and some related ones of similar size were on the northern slopes of the Witwatersrand, only a few miles from what one day would be recognized as the Main Reef, the world's richest gold deposit. Marais hurried to the capital, Potchefstroom, to report his news to the authorities and to petition for the right to exploit his discoveries. The result, in December 1853, was a most unusual contract between the Volksraad, or parliament, on the one hand and Marais on the other. The agreement held that he would be given a free hand to explore for gold and would report any discoveries to a newly appointed agent of the government, who was to travel with him for that purpose. Should any such find prove a commercial proposition, Marais was to receive a reward equivalent to £5,000 sterling. But should he 'make known any report concerning conditions on the gold mines that may be found or anything referring to these, to any foreign state, government, or individuals, by which the peace or liberty of this republic shall be imperiled,' he would, both parties agreed, be liable to execution.

Yet nothing untoward happened to Marais when his news, such as it was, did in fact leak out. He found no gold in payable quantities. The rigorously self-sufficient Boer government, with a sigh of relief as well as a tear for revenues that might have been, gradually lost interest in Marais and his proposals, and he slowly began to disappear from the view of history. But people had had their primal yearning for gold aroused, especially as news of California and Australia was still fresh in their minds, and in neighbouring jurisdictions a little less xenophobic than the Transvaal there was a movement to exhort a gold rush into existence. Businessmen in Durban, the principal city of Natal, began offering large rewards to whoever would prove up a goldfield there. Their enthusiasm was dampened, however, when a particularly promising nugget turned out to be of Australian origin, or so claimed the local chemist. Clearly, what was needed first of all was a higher level of geological expertise than could be found locally. In Cape Colony, a shrewd South African fresh from the diggings in Australia offered 'his services to any company now formed in this colony for the discovery of gold – the advertiser having a thorough idea of the nature and mode in which that precious metal may be found.' By 1859, the Boers were advertising as far afield as Germany for just such a person. In the 1860s, as depression set in on the Australian fields, prospectors turned to Africa with increasing frequency. Franchises were granted and mining companies were formed. But there was still a lack of capital with which to lure prospecting talent and to exploit whatever discoveries might result. The unexpected discovery of diamonds changed the situation fundamentally.

The story of an accidental discovery with great ramifications seems to us now an almost mandatory element of the lore of the rushes. It remains only to fill in the dates and names in each case. Accounts varied considerably and would become murkier with time, but it is clear that in June 1868, on a farm at Hopetown in the northern part of Cape Colony, not far from the border with the Orange Free State, either Erasmus Stephanus Jacobs or his young son or a farmer named Schalk Van Niekerk found what proved to be a 21¼-carat diamond lying on the ground. The stone was passed through several sets of hands in the search for someone who could render a learned opinion, and in time it was bought by the governor, Sir Philip Wodehouse. He paid £500, which, as it turned out, was more than the stone was worth, and exhibited it at the Paris International Exhibition. It attracted little positive attention. Everyone knew that diamonds came from Brazil or India and were invariably found in granite or mica, conditions that did not obtain in southern Africa. The diamond must therefore point to fraud: 'simply one of many schemes,' wrote one expert, 'for trying to promote the employment and expenditure of capital in searching for this precious substance in the colony.' Since then, of course, about one hundred tons of diamonds have been taken out of South Africa. Sir Philip Wodehouse's specimen is displayed as an almost sacred relic in the corporate museum of De Beers Consolidated Mines.

Despite what might be said in Paris or London, diamonds were materializing with ever-increasing frequency in southern Africa, and the pent-up energy characteristic of a great rush was starting to be felt. In March 1869, near the Orange River, a goatherd named Booi offered a 'pretty stone' to a farmer in exchange for a night's shelter but was turned away with the suggestion that he try Schalk Van Niekerk. He did so, presenting Van Niekerk with a bluish white stone of 83½ carats. It was eventually called the Star of South Africa. Van Niekerk paid Booi the equivalent of $2,000, making him by far the wealthiest herdsman in the region; then he quickly sold it for $60,000 to middlemen, who sold it in England to the Earl of Dudley for $125,000. This was merely the most sensational find of many, and word spread exponentially. An Irishman,

Captain Loftus Rolleston, led what he thought was the first diamond-hunting party to reach the fields but found that Australian fossickers had beaten him to the banks of the Vaal. The Aussies had the advantage, for many of them, displaced when capitalists supplanted the ordinary independent prospectors in the gold towns of Victoria, had come to Africa because it was closer than Europe or the Americas and, being less fully explored, held out the greatest potential. Another prospector who was quick off the mark in this new race for diamonds was Isaac (Ikey) Sonnenberg (1833–1906), who had left Florsheim in Germany, following the 1848 uprising, and headed for the California goldfields. He had served the Northern side in the American Civil War and then had come to Africa seeking gold. He eventually found it on the Rand but only after first taking part in the purblind rush for diamonds. He is thus a useful representative of a large phenomenon.

The parallels between the South African diamond rush and what had taken place in California and Australia are striking. The previously stable populations of the Orange Free State and neighbouring Cape Colony and Natal were thrown into social and economic chaos by the rush to what were called the Dry Diggings in the valley of the Vaal. Masters of ships lying at Cape Town and Port Elizabeth suddenly found themselves with too few hands. Soldiers failed to return from leave. Where a year before there had been only ten Europeans living along a hundred-mile stretch of river, now there were ten thousand. A year or two later, when excitement shifted from the Dry Diggings to a spot called Colesburg Koppie, one observer estimated that there were ten thousand people working an area only about twelve acres in size.

The majority of the miners were British, at least nominally, even after the news reached the United States in August 1870 with predictable and immediate results. It seemed to observers that every race was represented. It was just as easy to see that every class was present too. Clothing ranged from the elegant and entirely impractical to the plain sturdy homespun that had evolved as the unofficial uniform of the gold rushes. Many belts held revolvers or bowie knives, and this too was the product of experience.

At its widest, the Vaal was about three hundred yards, and the valley was lined with koppies, or kopjes – small hills. As in the previous gold rush sites, the climate was generally healthful, but it had the same wild swings between aridness and deluge which the forty-niners had experienced. Merchants throve because despite an abundance of antelope, there was little expectation that so many people could live off the land.

The camps that were springing up were more Australian than American in appearance, a combination of canvas and corrugated-iron sheets. Conditions grew less sanitary and the water supply became tainted and brought typhus, known locally as camp fever.

The same type of entrepreneurs who had flourished in gold rushes prospered equally well here, and in at least one famous instance the same entrepreneur – Freeman Cobb, the American stagecoach magnate – won a large share of the overland transportation market just as he had done in Australia and New Zealand, though against much stiffer competition. Another entrepreneur with a vision was Jerome Babe (fl. 1860–75), the Cape Colony sales representative of the Remington firearms company of America. He saw miners with U.S. and Australian experience hunting for diamonds with the same techniques and equipment they had used to search for nuggets, for indeed in its early stages the diamond rush was concerned with alluvial stones. Observing long toms rocking forward and backward as they sifted gravel, he conceived the idea of a machine that would accomplish the same task by a rotary motion and be smaller, cheaper, and more efficient. The device was named after its inventor, but Babe was corrupted to 'baby' as it entered the common language.

The lone miner quickly came to seem obsolete, however, after the discovery made in 1870 at Colesburg Koppie, a low reddish brown hill, thirty feet high, more than six hundred broad, on the north side of the river in a district dominated by thorn trees. A pipe, or deep volcanic fissure, was found to be rich in diamonds, and this gave hope that the free diamonds now getting scarce were only the beginning – that figuratively as well as literally the miners had only scratched the surface. The result was called the New Rush, which was bigger than the first, with twenty thousand people taking part in 1871 and sixty thousand by 1874.

One English observer urged the participation of any countryman who was 'young, active, strong, "smart," and above all, steady ... with a few hundred to spare' for tools, supplies, and thirty shillings per month for each of four black workers to do the heavy labour for him. For at first the local 'Coloureds' and the black labourers who were brought in from as far away as Mozambique were prevented by the Diamond Diggers' Protection Society from holding mining licences themselves. There were riots when this was later reversed by the British. Despite the reform, the authorities still maintained strict control over Africans, even to the point of requiring travel passes of the sort that later became so central a symbol of modern South Africa.[1]

Such politics were the result of a geological imperative, one that did not augur well for the independent white miner either. As the holes became deeper, it became more expensive to dig them, so the companies owning them grew bigger. The individual digger might experience the luck described in a folk song: 'A turn must come some day / a ninety-carat would change the past / and make the future gay.' The odds were pretty much what they had been in Australia. But most digger families (for this quickly became a more family-oriented rush than previous ones) found only drudgery, to say nothing of the scurvy, dysentery, and colic that were rampant in summer. All the while, immense economic forces were pressing down on them as small holdings were consolidated. As time went by, the business was dominated by fewer and fewer men, who grew richer and richer and brought in ever more black labourers, who were thus assured the dislike of the diggers as well as harsh treatment by the large companies. Capitalists knew that they could reduce the chances of a revolt by using Africans who spoke many different languages and dialects. Although there were almost no Africans on the Rand when diamonds were discovered in the 1870s, there were 15,000 there in 1889; the number shot up to 35,000 in 1892, to 42,500 in 1894, and to 51,000 in 1895.

The new field was three miles wide. It lay so near the frontier between the Cape and the Orange Free State that the question of ownership was moot. The Transvaal further complicated matters when the forceful figure of Paul Kruger (1825–1904), who had risen to power there in a series of small civil wars in the mid-1860s, claimed Boer sovereignty over the riches. By rights, the strongest argument was probably that of the Griquas, a mixed-race people who had the backing of the British government. There was also a vocal group of dissident miners led by a former able-bodied seaman in the Royal Navy, one Stafford Parker (1833–1915), who proclaimed the existence of a new Diamond Fields Republic with himself as its president. In what came to be known as the Black Flag Rebellion, he was aided by Alfred Aylward (1841–89) né Murphy, a Fenian rebel from the United States whose motives appear to have been more anti-British than pro-independence. The Cape governor, Sir Henry Barkly (1815–98), had previously served in Australia. Perhaps mindful of the disaster the governor of Victoria had made of a similar predicament, he nipped the matter in the bud in the traditional way, with a show – and only a show – of muskets and scarlet tunics.

In the end, the district was hived off as Griqualand West, a separate colony in close association with the Cape. The Orange Free State received a token payment of £90,000, the equivalent of a few days' take from the mines; the smaller diamond areas that were later opened up within its borders must have been a salve. Transvaal received naught, a fact that did not improve the disposition of Kruger, a rigidly conservative, dour, and warlike leader. But what was done was done, and the whirlwind of events swept through. The town that shot up in the middle of the vast diamond field was named Kimberley, after John Wodehouse, first Earl of Kimberley (1826–1902), the foreign secretary in London who was destined in 1881 to support the policy of home rule for the Boers following another period of British occupation. A small silver town in British Columbia was named after him as well, and later there was the Kimberley district of Western Australia with its own vivid gold rush associations. But Kimberley in South Africa has always been the one known to the world because of the fabulous stories surrounding its Big Hole and the equally fabulous characters who made their fortunes there.

The Big Hole was a craterlike aperture occupying the spot where Colesburg Koppie had stood. Mark Twain, seeing it in later years, reported that it was 'roomy enough to admit the Roman Coliseum.'[2] In fact, it was the biggest excavation ever made by man, an enormous open pit mine begun in response to the fact that slides kept closing off the cylindrical shafts that were dug first. To lessen the danger and to allow greater access to the band of diamond-bearing blue ground that lay beneath the so-called yellow ground, great benches, or earthen shelves, had to be cut into the pit's nearly vertical sides. The area was divided into hundreds of claims, most of them thirty feet square until they were cut into smaller and smaller pieces by the enterprising claim holders, who set up tents and canopies up and down the slippery slopes or simply stuck brightly coloured parasols in the mud. The work was carried out by the miners and their families, including young children, and by the thousands of male Africans brought in for that purpose. The term 'digger' no longer implied that one did the digging oneself.

At first, the earth from a claim was thrown by the shovelful from one terrace to the next until it reached the surface, where it was hauled away to be searched for stones. This method soon gave way to leather or metal buckets on pulleys, and wheelbarrows were replaced by large carts. There were so many wires for pulleys that the whole area, if viewed from above, would have resembled the fine-line cross-hatching of a steel engraving or a maddeningly complex spider's web. As the Big Hole neared the hundred-foot mark, the pulley system was replaced by horse or mule power, which was eventually replaced by steam engines. The hills of discarded earth on the surface grew large and grotesque, man-made koppies so to speak, in whose shadow was a confusion of tents and native kraals.

People who found diamonds in the earth certainly made money selling them, but the biggest profits awaited those who bought up one claim after another until they controlled sizable portions of the fields. And those who got in earliest naturally did best. Henrik Wilhem Struben (1840–1915) and his brother Fred (1851–1931) were among the most powerful diamond men of the time. Henrik later bought the farm of a Boer named G.A. de Beer. It proved so rich that the New Rush was for a time called de Beer's Field, though the name de Beers is usually associated with the mining company started by Cecil Rhodes (1853–1902), the most famous and perhaps most extraordinary figure of the South African rush and certainly the only one important for much more than his wealth, great though that became.

Rhodes was one of nine children of a Church of England vicar. His weak constitution made him long for adventure, while his subtle and cultivated mind inclined him towards scholarship. The two impulses were always in such conflict that it is difficult to say which of Africa's charms, its healthful climate or exotic atmosphere, brought him there in 1870 when he was seventeen. He joined his eldest brother, who was already established as a farmer in Natal, but when news came of the Kimberley fields he hurried there at once. A common story suggests that Rhodes carried with him only his mining tools, some volumes of the classics, and a Greek lexicon. He was a pale and slender fellow, for he had not yet assumed the blimpishness of figure that would later accentuate the blimpishness of his politics. At eighteen, he was making a hundred pounds a week in the diamond trade without exceptional effort. At nineteen, he suffered his first heart attack. The sudden awareness of his mortality must have permitted his other self to dominate, for he briefly abandoned Africa for studies at Oxford.

In truth, though, and despite his birth and childhood in England, Rhodes was a colonial, and as such he felt most awkwardly provincial when in the mother country and most thoroughly English only when in some faraway part of the empire. This allowed him to remain in Africa as the most fervent supporter of imperialism imaginable, one whose plan for a Cape-to-Cairo railway, for instance, was really no less than the waking portion of a dream, a dream of a totally British Africa. Only in that continent's isolation could Rhodes's patriotism flourish, because the distance allowed him to take cognizance only of what he found agreeable. No anguish about a free Ireland could infect him in Kimberley, where the temperatures commonly exceed one hundred degrees Fahrenheit and the flies buzzed around the trappings of imported opulence. Although of the generation that was agitated pro or con by the Fabians, he was untouched by their concerns, adhering instead to an older, more romantic vision of England. Of his friend Earl Grey he could say, 'Take heed of him, all of you, for in him you see one of the finest products of England ... an English gentleman.'

Kimberley diamonds were making such early capitalists rich indeed. Several nationalities were represented in the topmost stratum, as witness the important Compagnie française des mines de diamants du Cap, though the major players were usually British, if often by circuitous routes. For example, the admired and respected Alfred Beit (1853–1906) was born in Hamburg and came to Kimberley to work as a smous, or roving merchant, and by that means soon became an itinerant diamond buyer. Another smous turned big operator was Samuel (Sammy) Marks (1843–1920), who was born in Lithuania and arrived in South Africa by way of the Sheffield steel mills. Scores of such financiers performed a gaudy ballet of intrigue and collusion whose intricate movements were even then difficult to follow. The crucial year was 1880, when Rhodes formed De Beers Mining Company and his great rival, Barney Barnato, established Barnato Diamond Mining Company.

Barnato (1852–97) had a colourful past by the standards of the South Africa gold and diamond rushes, which tended to favour those with financial or engineering experience. Indeed, he would have stood out even in the earlier rushes, where improbable personal histories often seemed to be the norm. He was born Barnett Isaacs in Whitechapel, the grandson of a rabbi and the son of a publican. He assumed the name Barnato not to disguise his ethnicity (however resented they were, Jews were conspicuous leaders of the South African boom) but rather because it sounded more Mediterranean and so more suitable to a career on the stage. He was at various times an amateur boxer and a clown as well as an actor, both on his own and together with his brother Harry (1850–1908), with whom he washed up in Kimberley. The transition from cockney entertainer to koppie walloper, or diamond trader, may be difficult for the contemporary mind to grasp, but apparently it did not seem so extraordinary in context.

Such men as Rhodes and Barnato used heavy-handed methods when finesse seemed out of place. The theft of diamonds by African workers, for instance, was a problem they met squarely. Miners were paid as much as one pound sterling a day (approximately the same high wage by which Henry Ford revolutionized the American factory more than a generation later). Yet the figure was low considering the strain, the danger, and the way that any signs of trade unionism were met with violence. Many miners diverted some of the stones they found for later sale to IDBs – illicit diamond buyers – who toured the countryside, lurking outside the camps and in African eating houses, where they of course cheated the vendors outrageously. As the illegal trade amounted to hundreds of thousands of pounds a year, the mine owners did not hesitate to subject the workforce to sometimes brutal searches. It is recorded that twenty-eight stones valued at more than US$5,000 were recovered from a single miner after he had been forced to consume large quantities of castor oil. There is ample evidence, however, that diamonds were sometimes planted on suspects when genuine evidence proved elusive. In 1882, when Rhodes was already sitting in the Cape parliament (Barnato would follow), the colony passed an act that put the burden of proof on the accused in such matters rather than on the accuser. Finally, the De Beers company erected compounds where Zulus and other Africans were confined for their term of employment, living in shacks, cooking and taking their rest in a fenced and guarded common area, being searched frequently, and travelling to and from the mines each day by means of tunnels. The concentration camp is said to have been originated by the British a few years later, during the Boer War, but these compounds can truly be seen as a precedent.

6

The Rand and Western Australia

Just as performing artists can suddenly, after many years' work, find themselves famous 'overnight,' so could a country experience a gold rush after innumerable false alarms and near misses. The phenomenon had at least as much to do with psychology as with geology and was bound up with the effects of rumour on reality. Social, political, and economic conditions all had to be right before any amount of wishful thinking or any number of rich samples caused the gold rush crusaders to move en masse. Such was the case in the Transvaal. While the diamond-fields drama was being played out, the events that ultimately led to the Johannesburg gold rush were themselves moving forward. Kimberley and the Transvaal thus became two centres of energy, parallel but never equal in strength, the second being financed by the profits from the first once the proper momentum had been achieved. The process took time.

In 1868 a prospecting party scoured the Drakensberg Range, the high jagged mountain ridges in the extreme southeastern part of the continent. It had no success but seems important in retrospect. The members were Edward Button (died 1932), a Natal settler; Thomas Maclachlan (1840?–1900), a Scot; and two men known now only by their last names, Parsons and Sutherland. Parsons had previously been searching for gold around Durban, while Sutherland had a total of twenty years' experience in the goldfields of California, Australia, and New Zealand. Even then the word was being passed in the gold crusaders' community.

That such people were, for the most part, familiar only with alluvial gold delayed the widespread outbreak of gold fever, for the geology of South African goldfields was as different from that of North America or the South Pacific as the geology of its diamond fields was from Brazil's.

The gold in South Africa was in subterranean reefs, which stretched for miles and twisted and turned, often reaching great depths well beyond the scope of available methods and machines. They were such a difficult proposition for the gold-hunting technology of the day that the famous Lost Reef was not located until the 1930s, when it was discovered with the aid of magnetometers. Yet it was necessary for the early prospectors to continue searching if the political and economic conditions conducive to the big rush that people gradually came to expect were actually to come about.

In 1871 the ever-hopeful Button, by now with a different party in another area of the country, located a quartz reef and discovered its richly auriferous aspect. The report stirred some action in the Volksraad in Pretoria. No people wished more to be left alone than the Boers did – none made such a virtue of isolationism without introspection – but they were still smarting from the way they had been excluded from the wealth of Kimberley while their tiny nation underwent a serious financial crisis. As disagreeable as the prospect of an invasion of *uitlanders* (foreigners) might be, it was worth encouraging gold mining so long as the policy was not expensive and its results could be controlled. A gold commissioner was sent to Button's property, where he saw thirty people working a mine and various others (significantly, one from California and two from Maine) near by. He returned to Pretoria with rich samples.

Button went off to London to form a company to exploit the vein properly and sell shares. When he returned, he built a crude device he called the rocking boulder (now in the Geological Museum at Pretoria), a sort of teeter-totter, with black workers sitting on either end while others placed chunks of rock underneath to be crushed. By 1872, the machine had been replaced by a steam engine from Scotland. The government hurriedly passed a new set of laws that gave the mining interests maximum economic incentive while retaining the maximum amount of control.

Shortly afterwards the diamond miner Henrik Struben, having returned to Pretoria from the Kimberley region and purchased some farm properties with friends, was sitting in the Durban Club when he was startled to hear that gold had been discovered on his land, to which hundreds of men were flocking. A mining rush ensued with all the familiar characteristics, including the ability to ignite others easily, until the eastern Transvaal was experiencing some of the activity the Boers so wished for and so feared. This was all still alluvial gold and so in the long term a mere passade, but by 1874 an Australian, Henry Lewis, took gold by the same methods from a spot only forty-some miles from the future site of Johannesburg; in other words, from a reef that lay along an east-west line and extended sixty miles and would one day account for nearly half of the world's gold production – the soon to be famous Rand.

This was a gold rush in slow motion as well as in miniature. It featured several elements common in past excitements. As early as 1875, for instance, there was a diggers' protest meeting. Other developments were indicative of the future. The first government-appointed mining engineer was an Australian, while the engineers brought in by mining companies were almost invariably American; for such was the renown of Virginia City and Cripple Creek that British capital, in looking for engineering talent, no longer turned to Germany, the fount of such expertise in the nineteenth century, but to America.

Some diamond men, such as Ikey Sonnenberg, put money into speculative Transvaal companies, but at this stage the rewards were limited because of increasing political difficulties between the whites and Africans on the one hand and the British and Boers on the other. Only more gold, it seemed, could ease the tensions, by putting decisive power in the hands of one side or the other. All the while, the social pattern peculiar to the South African gold rush became increasingly clear.

When two small discoveries were made in 1881 and 1882 there was no question what to do. The first response was to register companies, issue shares, and bring in American engineers with hydraulic experience. The image of the lone prospector, with only himself and his pack animal for companionship, seemed remote, though some discoveries can have been on only a slightly larger scale. The Americans gave the place their characteristic stamp. Camps were suddenly christened Eureka and Fairview, and later there was one called Nevada. The price of tapping into the Americans' expertise was dealing with their temperament. One miner from California, after calling several times on the British nobleman who owned the company and failing to find him in, left a note that read, 'To Baron Grant, Your Lord Almightyness. It is beneath the dignity of an American citizen to call three times for his money. I have done so, and unless I get it at once, it would have been better if your mother had been barren.' The play on words was no doubt unintentional.

As the 1880s wore on, a consensus began to emerge in the mining community, a focus for its combined energy. With the spreading realization that South African geology was very different from what had been seen elsewhere, it became apparent that the task ahead was to locate the Main Reef, the primary plane of gold-bearing rock, which all were certain lurked somewhere beneath them. The Boers gave what encouragement they could by putting up a nominal cash reward. It was claimed in 1882 by Tom Maclachlan, the Scot who had travelled the Drakensberg mountains with Edward Button in 1868. There was a considerable flurry of staking, but an outbreak of malaria carried away a large number of those who had hurried to the spot. Many diggers looked disdainfully at the rush because it was not concerned with alluvial gold and promised employment without independence. This same reasoning, however, is what drew rather than repelled the capitalists and would-be capitalists. Struben retreated from the eastern part of the Transvaal to the western.

The seekers were like animals closing in on their prey. Each of a series of discoveries caused the pulse to race faster. A French engineer, Auguste Robert (1848–1908), who was known universally as French Bob, had gone to Kimberley after a varied career fighting both the Zulus and the British. Now he moved to the Transvaal and in 1883 located what he called the Pioneer Reef. Widespread chaos resulted, despite his efforts to keep the find secret. Then it was the turn of a former coal miner from Yorkshire, Edward Bray (1821–87). He had been exploring the Transvaal since 1860. One day, while preparing to throw a shovel at a black workman, he accidentally struck a rock, which broke open, revealing a streak of wealth his eye could not overlook. The assay showed that the rock contained eight ounces of gold to the ton. Bray's Golden Quarry was the name everyone gave the resulting enterprise.

Neither of these remarkable strikes was the by now almost fabled Main Reef, but the tension was mounting. In the Orange Free State, in Natal, and in the Cape, the level of expectation was as intense as in the Transvaal. In Africa, the word 'kaffir' was a noun, a derogatory term for Africans. In London, it became an adjective referring to South Africa in general, particularly so in the City, where 'kaffir shares' increasingly occupied the attention of speculators and investors. All the diamond tycoons were rushing to the Transvaal, especially to the central plateau. Herman Eckstein (1847–93), the German Lutheran who owned Phoenix Diamond Mining, was one. Others were Samuel Marks and Alfred Beit, and of course those bitter rivals, Rhodes and Barnato. The centre of gravity had now swung to the Witwatersrand. It must almost have seemed as though a pendulum or some giant plumb bob were hanging above the spot. Then, in 1886, came the great discovery everyone had been waiting for. In the manner of such events, it was somehow anticlimactic and almost comically accidental in tone, and the facts were instantly thrown open to dispute. Yet once it was acknowledged, southern Africa could never be even remotely the same.

In 1884 the Struben brothers began prospecting where the present-day uranium town of Krugersdorp stands. This was in the high veld, an area of rolling grassy hills, five thousand or so feet above sea level. In September they discovered what seemed to be a promising reef on a Boer farm a dozen miles west of what is now the heart of Johannesburg. They called it Confidence Reef, and it is an indication of their solvency as well as their confidence that they quickly began to work this and other properties on a large scale. Before whatever gold it contained could be extracted, the rock had to be crushed by stamping machines, mixed with mercury, and then heated to evaporate the mercury. It was a capital-intensive process when the ore, however abundant, was not so rich as that of California or Australia. The Strubens immediately had a battery of five stamps in operation on this latest discovery but came nearer and nearer the conclusion that the reef was not part of the Main Reef and probably was not worth the expense. They could not know that the Main Reef was about to be discovered by one or more of a trio of their low-level employees, who have come down in history as the Three Georges.

George Walker (1853–1924), a coal miner from Wigan in Lancashire, had gone to Kimberley in 1876 but had been unsuccessful and so turned to the cavalry, then to prospecting, then to the army again. At some point he had also picked up skills as a bricklayer and mason. He was back mining coal, this time in the Orange Free State, when he met an old acquaintance, George Harrison. Little is known about Harrison except that he was an Englishman who had been a digger in Australia and had had assorted scrapes with the law. The pair decided to travel together towards the Witwatersrand. As they approached, they heard that Fred Struben needed another miner at a place called Wilgespruit, and they vied for the job. Struben picked Walker and set him to work alongside a Cornishman named Arnold (whose role in the great discovery has been underplayed, some revisionists claim). Another man, Godfray Lys from Pretoria, joined them as well. While overseeing their batteries, the Strubens were boarding at a local farmhouse several miles from the site, and they ordered Walker to put up a cottage at a more convenient spot bordering a farm called Langlaagte, where Harrison soon ended up, working as a handyman. By South African standards it was a small farm, five thousand acres, and a poor one. It was also quite remote, a fifteen-hour ride from Pretoria on horseback. Another person on the cottage project was the third George, George Honeyball (1855–1949), a blacksmith with an English father and a Boer mother who was a distant relation of Langlaagte's owners. Up to this point, the various versions of the story are more or less in accord.

Walker later testified that on Sunday, 7 February 1886, while hiking across the veld to visit Harrison, he happened upon – in some accounts, literally tripped over – an interesting outcrop hidden in the tall grass. According to Honeyball's subsequent testimony, Walker brought the ore to the farmhouse, pulverized it between an iron frying pan and a discarded ploughshare, and panned the result in a nearby creek. He found unmistakable 'colour.' This topmost portion of what proved to be the fabulously wealthy Main Reef was conglomerate, pebbly in appearance, with the minerals widely scattered throughout the quartz. Locals later took to calling it *banet*, after a Boer confection in which almonds are sprinkled throughout toffee. The value of the formation was then unknown, however, and Walker thought no more about his find, he said, until several weeks afterwards when he was sacked from his job. He and Harrison then decided to go into business together; they signed a contract with the owner of Langlaagte to recover gold from the property, and applied to the government for a licence and for a proclamation designating the land a goldfield.

There was considerable bureaucracy implicit in all this, with President Kruger himself becoming involved. The government had been under increasing pressure to liberalize the mining regulations as a means of spurring exploration. The Boers were cautious, however, and in September 1886, before the red tape about Langlaagte had run its course, they announced that public digging licences for one hundred acres would be available at ten shillings a month for up to five years on lands proclaimed as goldfields. It was not the modest fee but the shortness of the term that provoked negative reaction, for mining in such country would be a lengthy proposition. In the meantime, rival versions of Walker's story had emerged, with Honeyball (who seems on the evidence to have been an honest and ingenuous fellow) claiming that he had made a similar discovery in a field of mealies (corn) the following day. Ignorant of mining, he had taken a sample first to Fred Struben, who had been somewhat abrupt, and later to Godfray Lys, who had looked at it and exclaimed, in a memorable phrase, 'By Jove, that's gold!'

There was a fury of negotiating all round. The government hesitated for one additional reason, the fact that the farm was set out for free grazing and was subject to an easement, known as a servitude in South African law. By the time all such matters had been resolved, the place was already aswirl with prospectors. Their unacknowledged leader was Colonel Ignatius Philip Ferreira (1840–1921), a Capetowner. Before the end of May he was already floating a share issue. Soon his name would be enshrined in a section of the instant city of Johannesburg.

Jo'burg, as sophisticates soon came to call it, was quite unlike any other mining town so far. It was a planned community, for one thing, though there were certain respects in which the planning went awry or was insufficient. It took permanent root almost at once, and not only grew by 6,000 per cent in its first year but also matured into a world capital with an economy still based on gold – unlike those of San Francisco or Melbourne. This is not to deny that the transition necessitated changes in the essential character of the place. Although in recent years the city has been associated with repression and social conservatism, it was actually quite a free and wild place in the early days.

In the beginning, the community had very narrow boundaries. Ferreiratown, some of whose flimsy houses had to be moved when it was discovered that they were too near the reef, was one of the few distinct areas. Another was Marshall Square, which was then a place set aside for the sale of horses and mules, though its name is now used like that of Scotland Yard, as a synonym for the criminal investigation department of the local police. Wishing both to accommodate the new economic demands and to preserve order, Kruger commanded a local surveyor, Josias De Villiers, to lay out a proper city. He chose a geometric design with a vast square at the centre. But the need to reroute a few streets around some existing claims spoiled the neat look of the plat. Had more been known of the configuration of the Main Reef, the town site might have been abandoned altogether, for today a wealth of gold lies unreachable beneath the high-rise canyons. The owners of saloons, or canteens as they were called, preferred corner locations, and their taxes were higher than those of other small businessmen. So the grid was laid out in short blocks to create an unusual number of intersections, a decision which rush-hour drivers of later generations would come to curse. The mining term 'stand,' meaning a campsite leased independently of the claim it adjoined, was used for the lots, with nearly a thousand stands in all, fifty feet wide and fifty or one hundred feet deep. At a three-day auction in December 1886, they were knocked down for figures ranging from £10 to £280. At one point, Barney Barnato actually attempted to buy control of the centre of the city, for speculation in raw land was practised at a feverish level, quite apart from the opening boom in construction.

The railway was three hundred miles away, and everything was dear. Wood at an exorbitant five pounds a cartload was used for cooking because coal was simply too expensive to bring in; even water was two shillings and sixpence a barrel, and of such dubious quality that it needed treatment with alum to make it transparent. Self-reliance, where possible, was a necessity. Many of the brick buildings, which rapidly displaced the initial tents and shacks, were the result of the industry that gave its name to the Brickfields, the city's first slum, and to judge by contemporary descriptions a nasty one indeed ('Let no one attempt a midnight exploration of the Brickfields without a lantern which is guaranteed to throw a light for yards distant,' the Johannesburg *Star* cautioned in 1890, 'otherwise the chances are that he will not leave the district alive').

One of the peculiarities of Jo'burg was that it had an instant aristocracy in the transplanted diamond tycoons, who were already beginning to be called randlords. This large wealthy class actually preceded a middle class, though until the proper setting could be erected its members could be seen only against middle-class backgrounds or worse. Cecil Rhodes, the richest man in town by far, sometimes sat, collarless, loafing on the *stoep* of the Central Hotel, a soft touch for beggars and panhandlers. Barney Barnato was to be found on many occasions playing billiards in the North Western Hotel at Pritchard and Fraser streets.

Social services were at an even higher premium than one might imagine. The Boers were powerless to stem the influx of foreigners, for in that freer age there were few non-excise barriers at frontiers. They could only extort high taxes and withhold Boer citizenship, two practices that rankled over the years. For a long time there were remarkably few paved streets for a city of such size and importance. The Boers had a particularly stout indifference to postal service, steadfastly refusing to initiate door-to-door delivery and erecting a pillar box – just one – only under the greatest pressure. In early Johannesburg, telegrams, once they had been received from the outside world at a station some distance away, were carried into the city by African runners. The chore of improving conditions fell to the randlords. Barney Barnato made a triumphant return to the stage in February 1889 playing the part of Bob Breirly in *A Ticket of Leave Man*, a play at the Globe Theatre in aid of the proposed synagogue. More to the point, he also built the city's waterworks.

If the city soon became an attractive one, with the low silhouettes and iron filigree already familiar to miners with Australian experience, it was a much more cruel and more violent place than Melbourne. Canteens were the most common type of business, and at first they served Africans as openly as whites – at least, the places of lower resort did – and often had free beds for the use of drunks. On Monday mornings, it was recorded, 30 or 40 per cent of black miners might fail to show up at their jobs. The bosses (or *baases*, as the whites expected the Africans to address them) decided to make up payrolls monthly rather than weekly in an attempt to slow the quick turnover in the labour force. Prostitutes, usually called 'sirens' in the newspapers, were certainly numerous but did not enter the local folklore the way they had done in San Francisco. Gambling was officially illegal but unofficially omnipresent. Street crime was a plague. From the press accounts, one forms a picture of a city where men routinely went about their business armed. One newspaper story spoke of knuckle-dusters becoming a common accessory.

Jan Christiaan Smuts, the future Boer leader and father of the Union of South Africa, called Johannesburg 'the Mecca of the hooligans,' and he was no promiscuous phrase maker. Rhodes's friend John Xavier Merriman (1841–1926), the former Kimberley diamond dealer and Cape Colony statesman, termed it 'the university of crime.' That it was indeed a rough place could be seen in its public entertainments. Dogfights were popular with spectators and gamblers. Newspaper editors (one of whom liked to carry a *sjambok*, a rhinoceros-hide whip, to discourage conversation) covered them like other sporting events. Ratting was another popular spectacle; a champion dog called Valentine was reported to have killed twenty rats in only two minutes, forty-three seconds. Boxing was still another. Barney Barnato was among the prominent figures who promoted the world heavyweight championship bout in 1889 between J.R. Couper of South Africa and Wolf Bendoff of England for a record purse of £4,000. Couper was declared the winner and wrote a novel before committing suicide in 1897; Bendoff remained in Jo'burg as a celebrated alcoholic. Even private fights between nonprofessionals were often relished as so much theatre. For years afterwards, enthusiasts contended that the most diverting match ever seen on the Rand was that which took place in 1893 between two miners named Shaw and Williams; at the conclusion of twenty-two rounds, Shaw lay unconscious and Williams required treatment in hospital. The most popular and important blood sport, however, was without question that practised on the Johannesburg Stock Exchange.

In no other mineral rush had the costs been so high or the need to form a capital pool so urgent. Yet in the early months of Rand gold, capitalists made do with the system that had been put in place to serve the diamond industry, most flotations being handled by the Kimberley Stock Exchange or by the Natal Stock Exchange in Durban. The need for a better arrangement was commonly acknowledged, particularly in view of the growing importance of British and European capital; and as early as February 1887, Joseph Benjamin (later Sir Joseph) Robinson (1840–1929) issued a prospectus calling for a combined gentlemen's club–stock exchange in the Rand. Robinson, who was often referred to out of his hearing as the Old Buccaneer, was probably second only to Rhodes in wealth. But he had no rival for the title of most distrusted and reviled randlord. When he died, the Johannesburg *Star*, no harsh critic of millionaires generally, published an obituary article under the heading *Nil neci maluum* ('Nothing but evil'). Few subscribed money to Robinson's venture, but by January 1886 an exchange building at Commissioner and Simmonds streets had been completed and was open for trading. The *Star* later called it 'perhaps the most hideous building that the perverted fancy of man ever imagined ... an eyesore to the nascent beauties of the town.' At one point, nearly a thousand brokers held seats in it. Barney Barnato got into a fistfight on the trading floor with the three brothers of the Lilienfeld mining family, and when he refused to apologize to the exchange governors, he resigned his place and bought the exchange and the stands on either side. By using workmen in three shifts, he had a new and larger building opened in November 1889, though it was pulled down in 1893 when still more space was required.

Only part of the action took place indoors. Shares too speculative to meet even the lax listing requirements of the board were bought and sold in a vast kerb market outside in Simmonds Street. Brokers and jobbers working in the exchange often ran out into the street to confer with their customers, who of course were not permitted in the trading hall. The authorities gave tacit approval to this over-the-counter market by stretching chains across Simmonds Street at Commissioner Street and Market Square, thus closing it to vehicular traffic. Later they went so far as to macadamize it, an expense not lightly undertaken in early Johannesburg. People who knew nothing else about the city knew about what went on 'between the chains.' Postcards with photographs of the greedy throng, a sea of bowlers in winter and boaters in summer, were virtually required of any tourist visiting from overseas. The buyers and sellers were frequently still going at midnight and even, to the particular horror of the Boers, on Sundays. As in Australia, buildings and land were commonly sold at public auction, and these events also took place between the chains. In 1896 a serious riot was barely averted when a crowd of two thousand, apparently disputing the worth of a suburban tract being put on the block by an estate agent, disrupted the share trading and threatened minor damage to the exchange building. Thereafter, the street activity was strictly policed and confined to a smaller area.

Throughout the first three years of the Rand boom a silent war raged between Rhodes and Barnato. In an escalation of the rivalry that had begun on the diamond field in 1880, each used charm, guile, and enormous sums of money to snatch competing companies from under the nose of the other with the aim of becoming the unchallenged master. Such was the game played for far higher stakes than ever before, and not even the indigenous wealth of these two fattest randlords could sustain the fight unaided. British and European families, such as the Barings and the Rothschilds, came to hold the balance of power. In the end, Rhodes won, not only because he enlisted Rothschild financial support but also because he struck Barnato in his area of greatest vulnerability.

It has been remarked, not without some truth perhaps but with a bewildering sense of proportion, that the attitude towards Africans that developed on the Rand had at least one point to recommend it: it used fuel that otherwise would have helped fire anti-Semitism. Because this diamond and gold rush was an opportunity for capitalists, not for mechanics and artisans, it was one that prized financial experience above mere craft. Thus, from the start, Jews were allowed a bigger role than they had been given to play in, for instance, California, where the most prominent Jewish gold rush figures were traders and merchants. The price of this importance, and the further wealth it engendered, was prejudice, which the Anglo-European world had not yet been taught to think of as a warning of something worse. The novelist Marie Corelli was simply giving voice to a standard sentiment when she complained after the Boer War that 'though the King is now "supreme Lord of the Transvaal," there is no chance whatever for British subjects to make fortunes there, the trades being swamped by Germans, and the mines controlled by Jews.'[1] To refer to Jo'burg as Jewburg seemed nothing more serious than a flippancy.

Many randlords took their wealth to London and attempted to cut a swath through society. As this was an age of plutocrats, all seeking respectability for their new money and themselves, many managed to realize their dream. Of all the colonials, however, the South Africans were set somewhat apart by the extreme depth of their pockets and what often seemed their irreducible enthusiasms. Sir Joseph Robinson, for example, continued to wear his pith helmet even in the West End. Jews such as Barnato were at a special disadvantage for which no freehold mansion in Park Lane could ever totally compensate.

It had preyed on Barnato for years that he had always been kept out of the Kimberley Club, an exclusive jockey club that occupied a none too prepossessing structure opposite Rhodes's house in the Kimberley high street. Barnato believed he had been blackballed because he was a Jew, and though there were in fact a few Jewish members, they may have been inducted in order to forestall precisely such criticism. Rhodes was at his shrewdest when in 1888 he offered to arrange for Barnato's election if Barnato sold out to De Beers, which would then change its name to De Beers Consolidated Mining. 'This is no mere money transaction,' said Rhodes. 'I propose to make a gentleman of you.' Rhodes authorized a cheque in the amount of £5,338,650, perhaps the largest ever encashed up to that time.

Just as Rhodes was executing the final move in his long game of consolidation, the whole gold rush economy came to an undignified halt. The collapse, for that is what it amounted to, began sometime in 1888 and was therefore unrelated to the panics that ruined so many lives and closed so many banks in the United States and Australia in the early 1890s. Its causes were wholly South African and were two in number. The first was the unrestrained level of sharp practice in the share markets, with insider trading, bucketing, and boiler-room operations pushed as far as primitive communications would allow, to say nothing of the incompetence that was rife in companies that were traded illegitimately. Too many small investors had been bilked too often; an entire sector of the market, one of great importance to continued exploration as well as to general confidence, simply withdrew. Lord Randolph Churchill made an inspection tour of the mining region and informed his readers in Britain:

In the early days of the Rand goldfield folly and fraud reigned supreme. The directors and managers were as a rule conspicuous for their ignorance on all matters of practical mining. The share market was their one and only consideration, the development and proper working of the mine being in many cases absolutely neglected. I was shown the Grahamstown Mine which, possessing only a claim and a quarter, was palmed off on the public with a capital of £120,000. This mine though situated on the Main Reef unfortunately struck a spot where the Reef was intersected by a thick dyke of clay, and it is scarcely an exaggeration to say that hardly an ounce of gold has rewarded, or will reward, the victimized share-holders. Millions of money have been literally thrown away. Bad machinery, badly put up, has been badly situated, badly worked.[2]

The bottom fell out of the market like the trapdoor on a gallows being sprung. All the larger gold-mining companies were interlisted on the London Stock Exchange, where their prices plummeted so much that, rather than suspend trading in them, the 'Change stayed open an extra day in order to bring its books up to date. On one especially poor afternoon in Johannesburg itself, a lot of 4,560 shares in fifteen different companies, many of them top performers only recently, sold for a mere £81. At one point, 169 brokers were on the verge of being 'hammered out' for failure to pay their membership fees. Mines even suspended operation totally. There was no activity around the headgear that dotted the landscape. Another characteristic sight on the Rand was the mine dumps, the mountains of sandy waste that were left after the ore had been crushed and the gold extracted (tumuli which 'compensate the city for the lack of any great architecture,' as H.V. Morton, the famous travel writer, described them years later).[3] Now they suddenly stopped growing, and this made them all the eerier.

The other cause of the recession was even more worrisome. It had to do with the same essential problems of engineering. What made the Rand so unusual and so rich was that the gold was stratified in Precambrian slate and quartzite, not found in veins in younger geological formations. In the Main Reef leader, the stratum of conglomerate containing gold was from a few inches to a few feet thick, while the Main Reef and the South Reef ran above and below, getting richer as they moved westward. In a three-dimensional rendering, they would resemble the tube lines on three levels at Paddington Station, except that each layer twisted and turned, rising almost to the surface, then plummeting out of sight before reappearing with different dimensions, even different characteristics. To mine such reefs, one began by trenching and then resorted to excavating a complicated series of variously named tunnels ('drives' were lateral, 'raises' went upwards, and 'winzes' went down). For all its danger and the increasing expense, the system was still working well enough at the time of the crash in share prices, but that was when a terrible discovery was made. At the depths now being reached, perhaps two hundred feet in some cases, the conglomerate gold was covered in pyrites, which meant that the amalgamation process using mercury was no longer workable or at least no longer profitable. Now the panic began in the souls of the randlords and quickly

spread outward to the investing public. Even Rhodes, a person not easily daunted by immutable facts of nature, grew bearish.

Yet it was not long before one of the powerful forces of the nineteenth
century, British ingenuity, began to provide relief. Three Glaswegians,
William and Robert Forrest, both medical doctors, and John MacArthur,
a chemist, had been experimenting with a new extraction process that
involved a weak solution of potassium cyanide (the same substance
with which jilted lovers took their lives in shame and despair, if the popular fiction of the day is to be believed). Their findings showed that this
solution when mixed with the conglomerate resulted in a black powder
which, when mixed in turn with other elements and heated, left pure
gold. In the case of coarser gold, mercury amalgamation could still be
used, with the cyanide process applied to the tailings.

The inventors had secured their first patent as early as 1887 and by
the spring of 1890 were conducting trials on the Rand; before the year
was out there was a plant fully operational on one of Buccaneer Robinson's properties. The new process arrived in the nick of time and so
boosted gold production that investor confidence returned with fresh
vigour, and by 1895 there was a boom greater than any before. But that
same year saw events that threw the whole region into war.

The purely political machinations of Cecil Rhodes and their remarkable consequences need only the briefest discussion here. His wariness
of the underfinanced and nationalistic Boers had hardened into weari-
ness; he considered that their continued rule would be a serious check
on business. His plan was to send an Anglo raiding party into Boer territory, believing that other people loyal to the Crown would rise up and
that the British government would intervene and then annex the Transvaal. Whether or not such a scheme would have worked, it certainly
failed miserably when Dr Leander Jameson (1853–1917) and his
mounted volunteers swept into the Transvaal too soon and without
logistical support and were easily captured. For such was the notorious
Jameson Raid. The uproar in Britain was so loud that Rhodes resigned
his position of the moment, the prime-ministership of Cape Colony, and
devoted himself to developing the vast new lands to the north, which he
had negotiated and wrested from the Matabele chief Lobengula (1830–
94) and then named after himself: Rhodesia. But he was soon back in the
thick of it, politically and militarily too, during the Boer War of 1899–
1902, in which Britain performed as he had wished it to but at enormous
cost. He was for a time besieged by the Boers at Kimberley, where it had
all begun.

Such were some of the news events that people in the Klondike, a world away, paid handsomely to hear or read about.

The leaders of Western Australia, a colony that seemed nearly as remote from the cities in the east of the continent as they themselves were from London, wished for a gold rush of their own to inject some life into the economy. As early as 1852 a reward had been posted, and ten years later the arch-goldfinder Edward Hargraves had been drawn there for a look, but he was no more successful the second time than James Marshall had been in the United States. The situation remained stagnant until 1886 and the excitement over gold finds in the Kimberley. When the news spread from that remote infertile region in the northern part of the colony, it drew prospectors from various places in New Zealand and Queensland (for one of the prerequisites of a gold rush was a substantial body of increasingly dissatisfied prospectors who were poised to move on, despite what precedent had taught them to expect).

The rush broke early in the year, and by June there were perhaps two thousand diggers on the site, having gone there via Perth and Fremantle. Few of them could have had experience of such country. There was little vegetation and less to drink; one group of two or three hundred diggers, it was reported, walked seven miles a day for water. There was the implied menace of the Aborigines, as there had been in Queensland, but the real danger came from the terrain in which only the Aborigines appeared able to survive. Many whites were felled by the heat. Swarms of flies and other insects threatened some with madness. Yet the gold was comparatively easy to locate, however laborious the gathering of it.

There were no towns as such because the Kimberley had no centre, even temporarily; if the gold was anywhere to be found it was everywhere. As the temperatures began to slacken in the afternoon, men wandered off in all directions, turning over rocks in their path to see whether they might be hiding gold – or might actually be gold. This activity was called specking, and to any detached observer it must have seemed a strange rite indeed. The other method was called dry-blowing, or panning off; the term 'dry-blower' was used for miners in Western Australia the way 'forty-niner' had been used in California and 'sourdough' was later used in the Klondike. Dry-blowing, a technique originated in Mexico, was simply panning for gold without water, lifting a dish full of desert dirt and sand above the head and swirling it about to let the wind carry off the lighter material, leaving the heavier to fall into a second dish at one's feet. It took some skill and was harder physical effort than panning. The partial mechanizing of the process was called dollying and consisted of shovelling the material into a container suspended from a tall tripod; when the hanging box was shaken, its contents, minus the dross, fell into a receptacle on the ground. When digging was required in Western Australia, the result was likely to be wet sand, which first had to be passed through a sieve or dried by fires in the excavations themselves.

In any case, there were no recognizable seams or leads of gold-bearing material, an absence that rankled with the New Zealanders even more than the lack of the plentiful water and lush vegetation they were used to. Conditions were so unpleasant that people remained only for as long as wealth was easily come by, and the wealth was never great enough to compensate them for the hardship. By February 1887, there were only about six hundred dry-blowers still in the Kimberley, and they were susceptible to any indication that small rushes were beginning elsewhere. In December there was a minor stampede to a sheep station east of Roebourne in the Pilbara region. Ironic tales of naifs who discovered gold through some mundane accident always heightened the sense of believability. In this instance, a boy named Jimmy Withnell was said to have found the first gold while throwing a rock at a crow. Later, a party financed jointly by the government and the local farmers found a quartz reef in the country east of Northam. This became the Southern Cross goldfield, for which many prospectors made haste, only to find that capital was in control and jobs were being offered rather than a new kind of freedom.

Beginning in the early 1890s, mining in Western Australia was always associated with Coolgardie and Kalgoorlie, towns in the south-central part of the colony whose names were so oddly euphonious that they lodged at once in the world's memory. Coolgardie, the smaller, came first, getting its start in June 1892 when two former Queensland prospectors, Arthur Bayley (1865–96) and William Ford, found a reef one hundred miles east of the Southern Cross. The two men took out much gold and thought they had the spot to themselves, but when they left to purchase more supplies they were followed back to the site of their triumph. The would-be claim jumpers neglected to observe the legal niceties, however, and Bayley and Ford were able to profit handsomely from their good fortune.

The fact that miners were on strike in the Southern Cross meant that there were many people free to hurry off to the new community of Coolgardie, but the attraction was far greater than that alone would suggest.

The news appealed to ordinary citizens in the way news of the most important gold strikes so often did, and in Perth and Fremantle every kind of shopbound middle-class person with no previous weakness for adventure left for the desert on short notice or no notice at all. It was a curious place they found, an instant city existing on water brought by camel caravans and often subject to rationing. An Irish-born miner named Edward Arthur Griffith sent word to Melbourne for his wife and children to join him. They travelled by coastal steamer to Adelaide and across the Great Australian Bight to Albany, thence by rail to Northam and Southern Cross, covering the last hundred and thirty miles by camel. One of the children later recalled, 'The sight that met our eyes when we finally reached the outskirts of Coolgardie and beheld the conglomeration of tent structures made of canvas and hessian was very discouraging indeed. The thought that this was to be our new home would have appalled the most resolute of pioneers.'[4] One of the most significant figures in the Coolgardie rush was the Honourable David Wynford Carnegie (1871–1900), second son of the Earl of Southesk (gold rushes being made to measure for second sons of noblemen). He was an important explorer and prospector, destined to die in Nigeria from a poisoned arrow.

One year after the founding of Coolgardie, Patrick Hannan (1843–1925), an Irishman with experience at Ballarat and in New Zealand, was travelling with two compatriots named Flanagan and Shea and discovered gold, it was said, only when he was forced to dismount because his horse had thrown a shoe. The three men staked what became known as the Golden Mile about two dozen miles northeast of Coolgardie; thus the birth of Kalgoorlie, which attracted the usual influx, both Other Siders from the east and miners from other parts of the world. This time they came equipped with condensers for 'cooking water' – for distilling potable water from a nearby salt lake. But here, as in Southern Cross, the nature of the deposits held out greater promise for stock promoters than for dry-blowers.

A further twelve months passed and another folkloric discovery was made. John Mills, a prospector returning with five friends from Coolgardie, stopped to rest about a dozen miles south of there and chanced to see a glimmer of yellow in a mossy rock that lay at his feet. The prospectors pledged themselves to secrecy; but one of them, it was said, let the truth slip out when he opened his mouth to take another drink. The result was a rush to what became known as the Londonderry Golden Hole. One side effect was a notorious swindle in which prospectors salted a mine and English investors were bilked of hundreds of thousands of pounds. But Londonderry, too, proved most conducive to large-scale exploitation, and not until more time had passed did the general run of dry-blowers get their last chance at independence, at Menzies, a hundred miles north of Coolgardie.

The rushes themselves were gruelling but not so cruel as some others. The miners at first played the part that history had written for them, banding together in committees until formal authority could be established. They found their own level of self-regulation, one that was more reminiscent of New Zealand than of California so far as violence was concerned. When a thief was caught, the proper response was a roll-up, in which someone beat on a pannikin to summon the community to watch the accused be banished from their midst. When government stepped in with a code of regulations derived from the Victoria experience, there was no great list of demands from an angry populace. There was not even a call for an official gold escort. Western Australia was too peaceful for that, even though every aspect of the development there seemed to proceed at breakneck speed, partly no doubt because of the way it was financed.

As in South Africa, the boom in the goldfields was mirrored by a boom on the share markets in London and elsewhere. Flotations with either Coolgardie or Kalgoorlie in their names were plentiful and assured of a certain speculative run-up. Between January and July 1895, 173 Western Australia mining stocks were given life in London, with many more on the Bourse in Paris and of course the Sydney and Melbourne exchanges. In fact, both Coolgardie and Kalgoorlie had exchanges of their own, where the action was fast and unpredictable, somewhat in the manner of Nevada and Colorado mining towns. The difference was that in Australia the ordinary prospectors took part, not simply as a system of insurance or a sign of faith, but as victims of a second kind of mania, quite apart from the one that had brought them there originally. Hundreds of thousands of shares changed hands when the two markets had 'open calls' each evening. Many miners became share gamblers because they were in debt to local shopkeepers, who quickly went beyond the role of grubstaker to become loan sharks or poor men's bankers.

In the two years ending in 1896, when the railway reached both cities, Western Australia gained seventy thousand population, though as late as 1897 the ratio of men to women was ten to one. It is not an exaggeration to say that the gold rush was a crucial factor in Western Australia's decision to join with the other colonies in forming the Commonwealth of Australia. One of the leaders of the federation movement in Western Australia was John (later Sir John) Kirwin (1869–1949), who arrived in 1895 and became editor of a famous newspaper, the *Kalgoorlie Miner*, before slipping into politics. The fact that the Kalgoorlie mines looked to have many productive years ahead of them attracted outside capitalists, so the town grew faster than its rival, and its professional middle class was bigger and more powerful. The promise of riches brought quick civilization accompanied by pride in a past that was rapidly receding. Today, there is a statue of Patrick Hannan in the high street of Kalgoorlie (population 2,300) but no equivalent in Coolgardie (population about 700).

7

Many Roads to Dawson

Although it drew tens of thousands of people, the Klondike stampede that crested in 1898 was not the most populous gold rush. Neither was it the wealthiest, though the gold, while it lasted, was plentiful and, compared with South Africa's, easily mined. In the popular imagination, however, the Klondike gold rush is the equal of the California one, indeed almost indistinguishable from it. To be sure, the forty-niner and the sourdough are the alpha and omega of the same alphabet. But it is neither the similarities alone nor the differences by themselves that best tell the story, but rather the pattern woven by the two together.

To those who had come to maturity in the 'days of forty-nine,' the United States of fifty years later must have been unfriendly and unrecognizable. Americans were using up the last of the western territories that had once seemed an almost inexhaustible resource. The nation was being forced to confront the very idea of a frontier (and consequently was increasing its adventurism abroad). The concept of a gold rush in the extreme northwest corner of the continent was instantly attractive. A young generation that had been reared in industrialized society was drawn to it as easily as older people, whose memories gave them the basis for a better informed, more empirical sort of hopefulness. The fact that it proved to be an individualistic affair, inflaming the imaginations of clerks and mechanics as well as of the floating subculture of hardened prospectors, made it irresistible. The country was filling up and settling down as the century was ending; there was an elusive element of millenarianism, the dark kind as well as the optimistic sort, in the way that news from the Klondike seized the attention of the masses.

There were engines of publicity that had not been available before, as well as great advances in transportation. Yet none of the other rushes had taken place in an area of such remoteness or harshness. An over-land wagon trip across the plains and deserts to California was a pleas-ant outing compared with some of the routes by which *cheechakos* (newcomers) scurried to the Klondike. People starved to death on gla-ciers and on high mountain trails, were killed in avalanches or drowned, or went mad or took their own lives in frustration at the impossibility of ever reaching their destination. By comparison with this arctic wilderness, Victoria of the 1850s was convenient to major centres of civilization. Thus the teasing, the tantalizing, the promise of wealth was all the more unbearable.

Although nearly as polyglot as the California rush, the Klondike was dominated by Americans to an extraordinary degree. It is estimated that 80 per cent of the people in Dawson at the high tide in 1898–9 were U.S. citizens. There is reason to suppose that a large number actually believed that they were in the United States when they were not; there is unlimited evidence that they behaved in accordance with such a mis-conception. The fact that they ran smack up against British law and Brit-ish administration caused tension that itself set the Klondike apart from other rushes. California had been run on the American plan of alternat-ing extremes, with individual lawlessness unrestrained until the vio-lence reached the level at which it was met with vigilante lawlessness. The British Columbia situation as personified by Judge Begbie and oth-ers suggested an alternative. Australia might have realized the social potential shown by British Columbia had it not been for a few such men as Sir Charles Hotham. The other rushes had been to varying degrees struggles to find a balance between the American and British styles, between levelling democracy and ennobling paternalism. The Klondike was slightly different, for it was not only the British colonial tradition but a set of distinctly Canadian values that stood in opposition to the rampant Americanism whose presence was necessary to define them.

On the Alaska side of the United States–Canada border were the lurid Wild West towns that were the staging areas for most of the Klondike argonauts (a word that had come back into vogue). Like the other Alas-kan centres that eventually usurped much of the Klondike's strength, they were vicious, deadly places. Sam (later Sir Samuel) Steele (1849–1919) – Steele of the Mounted – wrote of Skagway: 'At night the crash of bands, the shouts of "Murder!" and cries for help mingled with the cracked voices of the singers in the variety halls.'[1] The town was the eth-ical opposite of Dawson, over which he ruled with a just and steady hand. In Dawson, as in Johannesburg under the Boers, blue laws were strictly enforced. In complete contrast to Jo'burg, firearms were forbidden.

The Klondike's long-term effects on Canada were many. It opened up the North, or at least opened up Canadian eyes to the North's possibilities. It made important cities of Vancouver, Victoria, and Edmonton (just as it did of Seattle and other places in the United States). Perhaps most important of all, however, was the vivid, almost indelible contrast it gave between Canadians and their southern neighbours. No longer could Canadians too young to remember events such as the American Civil War continue to believe that what made America different was the simple absence of adult supervision. Now it was clear that there was a basic moral distinction.

Beyond the question of authority versus individual freedom, the Klondike phenomenon had all the characteristics of previous rushes, a cumulative set of traits and responses that had been gathering force for half a century. It was a rite of passage for thousands of people entering adulthood and a last fling at adventure for those at the other end of life. There were episodes that showed society at its best and at its worst: greed mixed with charity, prejudice with humanitarianism. Inventiveness went hand in hand with the deliberate rejection of all but the most primitive methods, just as ostentatious wealth coexisted with scenes of utter wretchedness without necessarily contradicting them.

Like other rushes, the Klondike seemed to fit the hole prepared for it in some giant unseen economic puzzle. Since 1870, the major European countries had adopted the gold standard one by one. The two major Asiatic powers, Japan and Russia, were soon to make the same transition. That left only China and a few Latin American countries on the silver standard. The United States, with its so-called free-silver debate, stood alone, agonizing over whether to coin unlimited amounts of silver, as the silver-producing states wished, or maintain a policy of bimetallism, with a strict ratio of silver to gold. The question was a searing national issue that led to the Wall Street panic of 1893, which solidified as depression. All the countries whose people traditionally went off to gold rushes were reeling from its effects. In Sydney alone, to take that example, thirteen banks suspended payment, and the colony's economy was thrown into chaos. The Klondike brought an end to the problem by stimulating services and trade, by making people's blood pump faster, and by suddenly dumping so much new gold into the worldwide kitty.

As in all previous rushes, the existence of the metal was known long before the fact became important enough, and conditions right, to trigger a mad dash by so many people. In fact, Yukon and Alaska had experienced a few homoeopathic gold rushes, purely local affairs, before there was any means of spreading news to the Outside, as the rest of the world was called.

In 1873 there was a short, sharp gold rush in the Cassiar Mountains of southern Yukon and northern British Columbia, centring on the Dease River. It was significant because it was a gold rush on tundra and so presented fresh problems for prospectors. The participants were British Columbians and the ubiquitous Americans who had crossed into the country via the Alaska panhandle. Once their job was done, many clambered back to American territory and went prospecting at Dyea. Two years later, a French-Canadian trader named Juneau and his partner, an American named Harris, ignited a small rush at the place that would afterwards bear both men's names in turn – Harrisburg, later called Juneau. At the same time, another Québécois, one Pierre Erussard, known as French Pete, found gold near by, which a San Francisco capitalist exploited with a stamp mill employing as many as three hundred men.

And so goes the story in-country as well, along the eastern half of the great Yukon River, which many people still believed flowed north into the Beaufort, not westward into the Bering Sea. In 1886 coarse gold was found in Fortymile Creek, a tributary so named because it was that distance downriver from the trading post at Fort Reliance. In 1892 there were similar doings at Sixtymile (sixty miles from the post but in the other direction). More followed in 1893 at Birch Creek, where a part-Russian worked the ramparts, or benches, above the Yukon with Native labour until Circle City and Fortymile became the centres of energy, with Birch Creek mainly a supply base.

Supplying the few thousand scattered prospectors was supremely difficult because the natural transport corridor, the river, was navigable only a small portion of the year. The task fell mainly to isolated agents of two large American-owned concerns, the Alaska Commercial Company and the upstart North American Trading and Transportation Company, which often grubstaked prospectors. Significantly, this system of advancing them provisions and supplies against a percentage of any future discoveries was one that, in temperate climates, came at a later stage of economic development, once the great rushes had passed. But then the only precedent for mining in the Far North was the Siberian experience, which did not provide many useful socio-economic models, only technical mining ones. Like the miners in Russia, those in Alaska and Yukon spent the winter burning off the overburden, often by as much as a foot a day, until they reached bedrock and could locate the putative streak of gold; the rock was then heaped in dumps beside the mine to be sluiced during the three warm months of nearly continuous sunlight.

The United States government had done practically nothing to develop Alaska since its purchase from the Russians a generation earlier, and there was even less economic activity in most of the adjacent parts of Canada. So remote were these regions and of so little concern to their respective governments that no one knew for certain just where the international boundary lay. A Dominion government surveyor, William Ogilvie (1846–1912), addressed the matter. The question was of scant concern to the Americans, however, until after they realized that the greater deposits of gold were on the Canadian side, and the Canadians were fully attentive only after it became apparent that American gold rushes brought American politics and American culture.

Circle City in Alaska and Fortymile in Canada were identical in their boisterousness. Unadulterated American frontier behaviour was allowed to flourish in the absence of any code of civilized conduct. Accordingly, they were both run by miners' committees, a form of government that was often the means of codifying bloodlust; and American-style mining claims were permitted in both places. Each such claim was from 100 to perhaps 450 yards of riverbed, from the crest of the hills on one bank to the crest of those on the other, and could be worked by proxy if the claim holder chose. Ogilvie at first recommended that Ottawa allow matters to proceed apace, but in 1894 Inspector Charles Constantine (died 1912) and twenty Mounties arrived at Fortymile to restore order. At Circle City on the other side of the Yukon basin, order was never restored because there had been no order in the first instance. Before there was a Dawson City and before there was a Skagway, the foundation for the stark contrast between them already existed.

In the Klondike, as in Australia and other places, there has been some controversy over who made the big discovery, the one that ended the series of small rushes and brought about the single gigantic event. The contenders were two whites, Robert Henderson (1857–1933) and George Washington Carmack (c. 1850–1922), and two Natives known as Skookum Jim (died 1916) and Tagish Charley, who were related to Carmack's Native wife, Kate. The question caused great bitterness between Henderson, a Canadian from Nova Scotia, and Carmack, the son of an

American forty-niner. Since then, it has sometimes become a matter of national pride for citizens of the two countries. One's natural sympathy for the underdog favours Henderson, but cold-hearted history inclines one towards Carmack, who contended that he made the find on 17 August 1896, a date still commemorated in Yukon with a public holiday.

Henderson was the classic gold crusader. He left home (his father was a lighthouse keeper) with the avowed intention of finding gold, a search that took him to Australia, New Zealand, and the American Rockies. In the North, he was the prospecting equivalent of a stock-market contrarian, deliberately working in some area other than the one to which a rush was under way. In 1896 he was on the Fortymile River at a place called Ogilvie, after the Dominion surveyor, where he met a trading-post operator named Joseph Ladue, an American who had taken part in a long litany of rushes, from the Mexican border to the Dakotas. Ladue had been in the North for fourteen years and had been one of the first people to scale the Chilkoot Pass, the narrow mountain path by which so many thousands would soon set foot in the fabled Yukon from the Alaska panhandle. Ladue's life as a trader had not prevented him from continuing to prospect for gold in nearby rivers and streams. One such was the Throndiuck, which seemed richer in salmon than in gold. The name of the river, which whites pronounced Klondike, was from a local Native phrase referring to the stakes driven into the water to hold their fishing nets. Yet Ladue remained a businessman, a booster in fact, and was known to grubstake wandering prospectors in the hope of generating economic activity. Henderson was one who accepted his offer.

Henderson had modest success in some of the streams, including one called Australia Creek (the name no doubt an indication of where some earlier searcher had served his apprenticeship). His big find, paying eight cents to the pan, was in a tributary of a stream that flowed into the Klondike not too many miles from where the Klondike joined the Yukon River. He called it Gold Bottom and, in the free-wheeling and generous manner of the early northern prospectors, told everyone he met of the discovery so that they could share the bounty. It was when returning from Ladue's trading post with another load of supplies that he met Carmack (who was sometimes called McCormick).

Carmack had arrived in the North by jumping ship at Juneau sometime before 1887, when William Ogilvie came across him at Dyea. He was a different yet very recognizable northern type, a would-be intellectual with a desire to go native. From his wife and others, he had learned the Tagish language as well as the Chilkoot pidgin by which the Native people and the whites communicated, and he was on friendly terms with the Sticks of the Alaskan interior. His good-natured personality probably aided him in his desire to fit into the Native culture, an ambition that his fellow Americans felt was no ambition at all. They looked down on him accordingly.

Henderson passed along his news about Gold Bottom, and Carmack acted on it, though not immediately and perhaps more from curiosity than excitement. He lingered a bit with Skookum Jim and Tagish Charley, but then, at the latter's urging, moved on to a parallel stream called Rabbit Creek. It was a place already sampled by Tagish Jim, who wished to be a prospector and join the white community as much as Carmack wished to be accepted by the Natives. Before leaving – and such is the crucial contention on which so much ill feeling later hung – Carmack promised to send word to Henderson if prospects on Rabbit Creek were good, in order to repay the kindness Henderson had shown him.

Carmack later claimed that it was he who found the large nugget in Rabbit Creek (soon to be renamed Bonanza Creek) on 17 August, though the two Natives maintained otherwise and said that Carmack had actually been asleep when Jim pulled it from the water. All of them instantly recognized the significance of the spot, from which they began taking gravel containing as much as four dollars in gold to the pan. They danced and smoked cigarettes in celebration. Soon Carmack had a shotgun shell full of gold. The next day they began to stake claims. As discoverer, Carmack was allowed two rimrock-to-rimrock claims of five hundred feet each; the others, one claim apiece. They set off to record them officially at Fortymile. Klondike claims were identified by the location of the plot upstream or downstream from the original claim on any river, so Skookum Jim's claim, the first upstream from Carmack's, became One Above, and Tagish Charley's, the other side of Carmack's double claim, became Two Below.

Moving downstream, the party encountered a group of four Nova Scotians who had come north from California. Carmack told them of his good fortune and urged them to stake their own claims on Rabbit Creek, which they did. At the mouth of the Klondike he met up with two French-Canadian prospectors and gave them the same tip. All these people were soon immensely wealthy as a result. Yet Carmack did not pass the word to poor Henderson.

The recorder of claims at Fortymile first mocked Carmack and so did most of the other whites. He was a known layabout, and the gold he carried in the empty shell was of a colour and texture which they had not seen in the North before. But gradually the truth penetrated and men began moving towards the Bonanza. Nor was the significance lost on Ladue. He quickly staked out a townsite at the junction of the Klondike and Yukon rivers, giving it the name of a Canadian geologist, George Dawson (1849–1901), and then opened a lumber mill for the building boom he so accurately foresaw.

Some of the early stakers sold their claims for a pittance in the first few weeks, either out of disillusionment or out of the need to return to Fortymile to purchase supplies; for the want of food became at once a great inhibiting factor. A look at those who were in fact able to stick it out through the initial period, and who therefore became wealthy, presents a wonderfully varied cast. There were Norwegians and Russians and Swedes (including one Charley Anderson, known forever after as the Lucky Swede); the Swedes somehow acquired the same reputation as the Chinese in other gold rushes, of diligently working the claims that were too poor for prouder people to bother with. One important early claimant was an Austrian, Antone Stander, who touched off more hysteria when he struck it rich on a stream soon called the Eldorado. Most of all, there were Americans of every description, from a former Tacoma barber to an ex-football star from Harvard to a one-time physical education instructor at the Seattle YMCA, who was soon a famous millionaire; they at once imbued the site with a certain lawlessness, or at least disputatiousness, and countered it with a flurry of miners' committees. Yet the most successful figure was a Nova Scotian, Big Alex McDonald, at least in the short term (for he died broke). He appeared to be little more than a lummox and was sometimes called the Big Moose from Antigonish, but no lack of social polish or verbal skill could hide his shrewdness or what became his mania for making deals. Dawson soon became a very important city and he one of its most powerful citizens.

For three weeks, Henderson remained unaware of what he had wrought. When he learned of what, in his view, was Carmack's treachery, he began staking his own stream before the horde invaded. Then he met a German named Andrew Hunker, an old veteran of the Cariboo rush, who had already staked a discoverer's claim near by that was even richer. Henderson decided that it would be prudent to abandon his own status as discoverer and stake the obviously far richer five hundred feet next to Hunker (who soon lent his name to what had been Gold Bottom Creek). A later claim Henderson made elsewhere, on Bear Creek, was ruled invalid because in his absence the mining laws had been refined and tightened in reaction to the heavy American influence. So he had a grudge against the Canadian government to carry alongside his grudge against Carmack for the rest of his life. A bitter man now, he tried to leave for Colorado but got stuck at Circle City, where he took ill and, in order to survive, was forced to sell his Hunker Creek claim for a mere $3,000. In later years, his son spent much time, money, and effort trying to locate the mother lode from which aeons of erosion had carried the gold that others were scooping up. But as George Woodcock observed, 'Mother lodes are to gold maniacs what mirages are to desert travellers.'[2]

So localized was the splendid chaos of Dawson that the people at Circle City, a mere 220 miles away, knew nothing of it for months until a lone Dawsonite arrived by dog team. The newcomer began to flash his gold in a saloon run by Harry Ash, a veteran of the Black Hills stampede of twenty years earlier. Ash promptly announced that he was off to the Klondike and left his place of business without bothering to lock up or collect from the assembled drinkers. In what seemed an instant, the price of a good sled-dog jumped tenfold to $250 as Circle City became a ghost town. Among those dashing to Dawson was a former Texas marshal, George Lewis (Tex) Rickard (1871–1929), who became a promoter of fights in Dawson and went on to own Madison Square Garden in New York. Also in the crowd were the first of a great many entrepreneurs, many of them women, who grew rich catering to the needs of the others. In a community that was still mostly tents despite the best efforts of Ladue's sawmill – and one that was on the verge of starvation until the arrival of a flatboat-load of cattle – the definition of what constituted luxury goods became absurd. Salt was exchanged for gold of equal weight, and one man paid $800 for a keg of used nails. Even simple services such as laundering clothes proved immensely lucrative.

William Ogilvie sent news of the Klondike to his superiors in Ottawa, who responded the following year with a pamphlet to which few paid much attention. Meanwhile, he carried out a survey of the townsite and then of the claims, which were in a state of confusion. Many had been staked by guesswork, and some had come open again after sixty days when the original claimants had given up. When correctly redrawn, many of the claims had spaces between them ranging from a few inches to a few yards, and these became of great significance when they separated two properties that were already known to be rich. A miner named Dick Lowe, who had followed the crusade from the Black Hills, through Idaho and Montana, acquired an irregularly shaped piece of land at the confluence of Bonanza and Eldorado that proved the wealthiest of all. He celebrated with a lifelong drinking spree. People with rich claims paid others to do their digging for them, usually at the rate of $15.00 a day (compared with $1.00 or $1.25 a day earned by a semi-skilled labourer Outside). Big Alex McDonald, who soon had interests in nearly thirty claims, began the practice of treating the claims as securities – things to be bartered, parsed, leased, and used as collateral. But in no sense, certainly not at this early stage, did the mining become big business as it had in South Africa, though strictly on the basis of numbers it was certainly business big enough.

Dawson – taken to include the adjacent community called Lousetown – occupied a low, muddy stretch of shoreline in the shadow of the great scarred mountain that soon became a famous geological landmark to many thousands around the world. In the spring of 1897, a year before the great rush, it was already a town of two thousand or more and had the character with which it would later be associated. The depression raged in the cities Outside, caused partly by a shortage of gold as more countries opted for the gold standard and fears grew that the United States would suspend foreign payments in gold; it was common for people to hoard gold coins, which consequently grew scarce and traded at a premium over paper currency. In Dawson, however, Gresham's Law was reversed, with good money driving out bad. Everyone paid everyone else in unrefined gold; bank notes were prized for their convenience but were so scarce that they changed hands at a premium of perhaps 20 per cent over their face value.

One of the local notables was William F. Gates (died 1935), known as Swiftwater Gates from when he operated a river ferry in Idaho during the Coeur d'Alene gold rush. He was among the first to become rich on a large scale from Klondike gold, and certain traits of his personality became almost official elements in the make-up of the stock Dawson type. He was a master, or so he must have thought, of the grand gesture. In one famous incident, a trollop on whom he doted arrived at a restaurant on the arm of a rival and ordered fresh eggs, the most expensive item on the bill of fare. Gates promptly bought up every egg in town and created a massive omelette, according to one version of the popular tale. Similarly, he was a great one for tossing his fat poke on the bar and standing rounds for the entire congregation. In 1897 he was still the only resident of the town with a starched collar for his shirt and was said to hide in bed on the infrequent days when it was being laundered rather than tarnish his image by appearing in the mud street or the bare-board saloons in a state of dishabille. He and another man conceived the idea of opening a saloon and dance hall called the Monte Carlo, soon to be Dawson's most storied, but his partner made the mistake of dispatching Gates on the long journey to San Francisco to purchase inventory and recruit chorus girls. It was a long, long wait for his return.

In the summer of 1897 the rest of the world learned what had been happening at Dawson and reacted even more strongly than it had when news had first come from California or Australia. It reacted indeed with something close to dementia. This seems to have been linked to the fact that the world was now even more heavily industrialized and countries were much more interdependent. Dawson's first generation of *nouveaux riches*, including Ladue, were eager to clear out with their treasure. As soon as conditions warranted, approximately eighty of them – Americans, Canadians, Europeans, and Australians – boarded two steamers, the *Portus B. Weare* and the *Alice*, for St Michael, the former Russian port on the Bering Sea, where connections could be made to U.S. points. The *Weare* carried so much gold that the decks had to be reinforced. The passengers and their extraordinary cargo then transferred to two decrepit steamers owned by the rival trading firms, the North American Trading Company's *Portland*, bound for Seattle, and later the Alaska Commercial Company's *Excelsior*, destined for San Francisco. Many of the millionaire passengers were still dressed in their mining clothes, now in tatters. They gorged themselves on fruit and vegetables, whose taste they had almost forgotten.

The *Excelsior* had the longer journey but was the faster ship. Its arrival created such a stir that the San Francisco mint was closed temporarily, and a throng of spectators pressed against the private smelting firm where the gold was weighed and exchanged for currency, with joyous pandemonium the result. It is a statement on how rapidly the news grew in importance that two days afterwards the *Portland* was met by five thousand cheering people when it tied up in Seattle well before dawn. In its three extras that day, the Seattle *Post-Intelligencer* reported that the vessel carried 'a ton of gold.' In fact, the gold was two tons but that didn't matter; the phrase worked like smelling salts to revive a sluggish economy, though the price was disruption in the short term.

The Klondikers were celebrities and were hounded as such. Ladue, termed in the press the 'mayor of Dawson City' and 'the Barney Barnato of the Klondike,' fled eastward across the country but was harassed by his own sudden fame at every stop along the way. Few seemed to have a true sense of where the Klondike was, but the name was almost instantly part of everyone's vocabulary, though in the early days it was often spelled Klondyke and for years continued to be rendered as either Klondike or Klondyke. It is difficult to say to what extent the press was the engine of mass hysteria and to what extent merely its instrument. William Randolph Hearst, who built his newspaper empire on an inherited fortune that his father had made in another rush and knew something of the chemistry involved, worked hard to manipulate the situation after first being scooped in both San Francisco and New York. Those with direct past experience of gold rushes were also quick to see the import of what was happening. W.S. Stratton of Cripple Creek began buying riverboats suitable for Yukon navigation. E.J. (Lucky) Baldwin, a forty-niner and now a wealthy San Francisco businessman, announced that he was leaving for Dawson though he was seventy-one.

The Seattle *Times* watched as most of its editorial staff joined the gold rush. The city's streetcar system came to a halt as its conductors and others quit their jobs, as did police officers and even the city's mayor, who heard the news while on business in San Francisco and raised capital to buy a ship without bothering to return to his duties or even go home to pack. About fifteen hundred residents of the city left in the first ten days, and ships were lined up to take more. Medical students accelerated or abandoned their studies. Chicago gamblers saw the opportunity of a lifetime. Many businesses, from California orchards to Alaskan canneries, suffered for want of a labour force or because of the collapse in transportation.

Five steamers a week left Pacific ports for Alaska, crowded (invariably, overcrowded) with past miners and would-be miners. But the hysteria was not confined to the West and Northwest or even to North America. In February the first shipload of Australian prospectors left Circular Quay in Sydney, and Europeans followed in some numbers. Meanwhile, the get-rich-quick nature of the dream, coupled with a misunderstanding about which country the Klondike belonged to and how far away it was, made the adventure appeal to Americans in particular, who responded in typically American ways reminiscent of 1849. 'Hurrah for the Klondike!' became as popular as 'Ho! for California' had been. The classified-advertising sections of American newspapers were filled with appeals from people seeking grubstakes or Klondike companions, or investors to enter into syndicates in the Far North. Display advertisements were full of camping gear and dehydrated foods and practically any other goods in connection with which the word Klondike could be used profitably.

Some schemes were majestically far-fetched. A number of people, including a man in Dublin, announced plans to travel to the Klondike by hot air balloon. A great many seemed to believe they could go there by train or even bicycle. The poor benighted fools had no conception of how arduous a journey it was and how many would be forced back or would die along the way. It is estimated that in 1898 and 1899, one million made plans for the Klondike and perhaps one hundred thousand actually set out, with perhaps forty thousand making it to Dawson and maybe one per cent of those striking it rich. But the Klondike, even more than other gold rushes, brought out the gambler trapped in people's souls. The odds against them counted for very little in the excitement of being dealt the first hand.

A number of people did not survive the initial leg of the journey, the voyage up the Pacific to the Alaska panhandle, and many survived it only barely. Sixty-five died, for example, when one ship, carrying dynamite as well as passengers, blew apart between Skagway, the town then abuilding on the coastal mud flats, and Juneau. There were several schemes, considered progressive at the time, for sending single and widowed working-class women to the Klondike to find matrimony or self-fulfilment. In one of them, the Women's Clondyke Expedition, five hundred women left New York by steamer and – shades of 1849 – were briefly shipwrecked after rounding the Horn; only one ever reached the Klondike. The sidewheeler *Eliza Anderson*, a veteran of the Coeur d'Alene gold rush and more recently a floating casino in Seattle harbour, had a number of mishaps and ran so low on fuel that passengers had to chop up the furniture and the stateroom walls to feed the boiler. The ship actually made it as far as St Michael, seventeen hundred miles from the goldfields, whereupon the argonauts chartered another vessel for Skagway. Their difficulties were just beginning.

From Dyea Inlet on the coast of Alaska, there were two routes through the Coast Mountains into Canadian territory and upriver to Dawson. From the tent and green-timber settlement at Dyea one could take the thirty-five-mile Chilkoot Trail long known to traders and prospectors, emerging at Lake Bennett by way of Lake Lindeman. It was a fairly direct route and given normal weather conditions was passable most of the year even though it was high – too high for pack animals. The new alternative route was the White Trail beginning at Skagway, three miles from Dyea. Although it was forty-five miles long, it looked promising because it was six hundred feet lower in elevation at its highest point and hence was viable for beasts of burden. The first few miles consisted of a level wagon road. It sounded almost inviting to the inexperienced, but it proved a hellish place even for those inured to hardship and disappointment by having survived the different Hades that was Skagway.

The John Sutter of Alaska was Captain William Moore (1822–1910?), a German who for fifty years had been following gold rushes up and down the western edge of the Americas, from the Fraser, the Cariboo, and the Cassiars in the north to Peru in the south, with California the magnetic centre of his dreams. He made and lost a fortune in the steamboat business during the Cassiars rush and pushed on to Alaska in 1887, where one of his sons told him of the White Pass, which he hired Carmack's friend Skookum Jim to help him survey. Long before Carmack's discovery, he sensed that the gold in the earth's crust along the rim of the Pacific must extend to the territory beyond the Coast Range; he was determined to benefit from the rush he knew would take place one day. Yet he also feared the despoliation of his personal paradise, a lone cabin on the shores of the Lynn Canal, surrounded by cold placid mountains, remote from civilization. It was as though he were willing the gold rush to take place while convincing himself that he would limit or at least control its destructive energy, which of course he could not do; no one could.

In late July 1897 the first steamer full of gold seekers put in and disgorged a mob that promptly began to erect a jerry-built town – the town of Skagway. Ship followed ship, and in the American manner a rough attempt at government took shape. In August a miners' committee forced Moore off his property without compensation, laid out a mud street called Broadway, and discovered a thief, a Frenchman, who was tied to a stake and riddled with bullets, his body left hanging as a warning to others. Moore never quite gave in to the mob, but he had his revenge by building a mile-long wharf out over the mud flats and extracting a fee for its use from the captains of incoming ships.

Skagway was a world of its own and a law unto itself. Alaska was legally dry, but the town paid no heed and saloons with names such as the Mangy Dog, the Home of Hooch, and the Palace of Delight operated round the clock. The social elite was dominated by pawnbrokers and faro dealers, with blackjack boosters, or shills, at the top; the middle was occupied by bartenders and by prostitutes with names such as Ethel the Moose and Mollie Fewclothes. An Italian sold balloons in the mud streets, while a Russian led a dancing bear on a leash. But if there was a carnival atmosphere at times, it was that of a deadly carnival indeed.

The leading citizen was Jefferson Randolph (Soapy) Smith (1860–98), who ruled the town as surely as Al Capone later ruled Chicago. As a young cowboy in Texas, Smith had fallen victim to a con man named Clubfoot Hall and had resolved to learn all the subtleties of fraud himself. A bunko artist named Taylor taught him the patter for selling bars of soap to crowds of people who had been led to believe that twenty-dollar notes were hidden beneath the wrappers; hence his nickname. In Leadville, Colorado, he was the leading employer of con men and a figure, it was thought, with some flair. He once staged a road race between two naked prostitutes with a quart of whisky as the prize. He took over the nearby silver town of Creede, fleecing the miners in a dozen different ways and shooting or having shot those who objected too loudly. Now he brought all he had learned to Skagway, that paradise of credulity where hopefuls arrived with pockets full of cash and through which some returned with pockets full of gold.

Smith's brazenness extended well beyond watered whisky and crooked roulette wheels. He ran a phoney freight-packing company, several phoney information bureaux, and a phoney ticket office. Friendly strangers, including one dressed as a clergyman, met the *cheechakos* at the boats, won their confidence, and steered them to the various bogus businesses. The boldest was Smith's telegraph office, where newcomers waited in queues to send five-dollar messages to the States. (Skagway, of course, was not connected to the Outside by telegraph.) The same naifs dug into their pockets again when replies, written by Smith's large staff of con men, arrived collect. If these and other methods failed to separate an argonaut from his capital, there was always simple robbery and even murder. Smith appointed and paid the police force himself. He also performed marriages and granted divorces. When a legitimate churchman complained of all the crime, Smith contributed $1,000 to start a fund for cleaning up the town; the crusading minister was robbed of the money once contributions reached $36,000. In the end, this frontier community responded with the only device it knew or cared about, a vigilante group called the Committee of 101 – the same name as that used in San Francisco – and Smith was gunned down in public in July 1898 by a man named Frank Reid who was mortally wounded in the exchange.

An estimated five thousand people left Skagway by the White Pass route in 1897, but hardly any of them made it to Dawson before winter descended like a leaden curtain. The problem, basically, was one of supply. Cities and towns throughout the Northwest, on both sides of the

U.S.–Canada border, were beginning to grow rich from the gold rush trade, a trade that influenced their economies for years to come. Six decades later one could still find brand-new gold pans, stacked in their original packing grease, on the shelves of a certain ironmonger's shop in Victoria, British Columbia. Outfits bought in Canada had to be carried across U.S. soil in bond; they could not be opened in Skagway, yet while there they had to be protected from the thirty-foot tides as well as from bands of thieves, who could be discouraged only at gunpoint. Alternatively, one could buy all one needed in Skagway or Dyea, but at much higher prices, and then pay duty at the North-West Mounted Police encampment at the summit of the pass, assuming that one got that far.

Travellers on the White Pass, usually with their supplies on pack-horses, ascended from swampland to a treacherous winding path only two feet wide, then up a narrow alley of high rock, followed by a thousand-foot climb up a muddy mountain laced with sink-holes into which horses simply disappeared. Two further mountains blocked the way between the border post and the beach of Lake Bennett. Perhaps as many as three thousand horses died on the White Pass in 1897, whether by falling, being beaten to death, being left to die after collapsing of exhaustion, or being shot by the Mounties if judged too seriously injured to continue. The memories that many people took away from the gold rush included the smell of rotting horseflesh and the sight of half-frozen carcasses with vultures circling above. Even hardened observers such as Major James Morrow Walsh (1843–1905) of the North-West Mounted Police were struck by the inhumanity and suffering they saw there.

Anything that would lessen the ordeal or make the goal more plausible was welcome and therefore lucrative. Packers were highly paid and much in demand. One of the most notable was a poor widow, Mrs Harriet Pullen, who made one fortune during the day with a freight wagon and four horses, and another at night baking dried-apple pies. The former mayor of Minneapolis, ruined in the 1893 panic, built a wooden toll road along one stretch of mountain and profited handsomely. But there was still little alternative to packing supplies and equipment, one load at a time, one after another, while snows as deep as twelve feet covered over the main body of crates and boxes. One extraordinary couple actually managed to move two small sternwheel steamers over the mountains in pieces in the winter of 1897–8. One of the curious sights must have been E.A. Hegg (1867–1946), a photographer from Washington State, who packed his cameras and darkroom on a sled pulled by goats and who left the most complete and remarkable photographic record of the Klondike rush.

Tappan Adney (1868–1950), the *Harper's Illustrated Weekly* correspondent and the author of one of the most important first-hand accounts of the Klondike, swore that one horse was seen to commit suicide. This was not the only such alleged incident (another involved an ox). Jack London crossed the White Pass but later exaggerated his exploits about running the rapids that followed. Despite his melodramatic prose, he caught what must have been the flavour of the experience in one of the many Yukon fictions by which he made his fortune. He wrote, 'Their hearts turned to stone – those which did not break – and they became beasts, the men on the Dead Horse Trail.'[3]

Although the Chilkoot Pass was no stroll in Hyde Park, neither was it a wrestle with death, as the White Pass proved to be. Nor did Dyea present the same dangers as Skagway, despite the presence of Soapy Smith's agents there too. Through word of mouth, the Chilkoot became the most popular means of access to the Klondike. There is no reliable way of determining how many of the people who began the journey by this route backed out, though many certainly did, for the Canadian authorities turned away some, and the frustration level was sufficiently high that a number of suicide cases were reported. Yet fully twenty-two thousand made it to the Chilkoot summit and over.

The image of them in photograph after photograph, a thin black line of figures made to seem ever more antlike by the size of the mountains and glaciers, has become one of the most memorable in Canadian history, as well as being one that is instantly recognizable in many other cultures. It is significant that Charlie Chaplin used the Chilkoot Pass scene and carefully identified it as such in a title card at the start of *The Gold Rush* in 1925. The film had even less documentary flavour than most other movies about the Klondike, so wanton in their disregard of historical accuracy.[4] But Chaplin knew that by showing the prospectors climbing in single file and hand over hand, he would conjure up a wealth of emotional responses. Even a quarter-century after the event, audiences around the world knew what the Chilkoot Pass represented and what determination was entailed in crossing it.

Dyea had but a brief existence. In 1897 it consisted of a single trading post. In 1899 it was rapidly disappearing, made obsolete by the opening of the railway linking Skagway with the new community of Whitehorse on the Canadian side. During the intervening year, however, it had a population of three thousand, though never the same three thousand from one day to the next, and presented a blur of activity as people struggled to get their goods ashore and safely above the tideline. Wharves and warehouses were soon erected, but these were signs of social decay, not of social progress, for by then the miners' code was breaking down, and one could no longer leave supplies untended. Wilson Mizner, the future playwright and eventual owner of the Brown Derby restaurant in Hollywood, who was in the Klondike with his equally colourful brothers Addison and William, recalled the arrest and 'trial' of three accused thieves along the Chilkoot. One broke away and shot himself rather than be recaptured. A miner volunteered to give one of the others fifty lashes, a task he carried out with obvious psychosexual delight. The story could only belong to 1898, because in the smaller, less desperate days of 1897, theft among the miners was almost unknown, and people boasted of never locking their doors or hiding their possessions.

For several miles, the trail leading from Dyea was a passable enough wagon route, but it soon became difficult, as witness the mounds of abandoned goods on either side. These included rubber boots, which a pair of Americans collected for resale in Juneau. After a corduroy toll bridge across the Dyea River, which the miners soon overran and opened to the public, there was a treacherous canyon, two miles long and only fifty feet wide, followed by first one camp and then another. The second of these was Sheep Camp. Because it was the last opportunity to gather firewood, hunters going after mountain sheep had long used it as a base. Here there might be a thousand people availing themselves of a wide range of facilities, from laundries and grocery stores to whorehouses and barber-shops. Here too there were private post offices, a response to the government's maddening inability to run an efficient mail system in gold rush areas, even Dawson. It was at Sheep Camp, with its population of exotic transients, including a group of Maoris from New Zealand, that horses had to be abandoned, and thousands were shot or left to starve. When the fog allowed, one could look up and see the Chilkoot Pass four miles away, a small gap, five hundred feet lower than the peaks on either side of it, like a broken tooth in the middle of a jagged comb.

At the top of the pass a tattered Union Jack beat time in the wind, indicating the North-West Mounted Police station and the customs shed where Canadian authorities screened the successful climbers and collected duty on their outfits (to the extent that $150,000 in cash had to be shipped out via Skagway under the noses of the Smith mob). Their presence at this strategic but lonely spot, where one snowstorm raged for two months without interruption, was due to an international diplomatic problem. Britain and Canada differed with the United States on interpretation of the boundary agreed to by the British and Russians as long ago as 1825, more than forty years before the United States bought Alaska from Russia. The Americans believed that their territory extended inland from the panhandle coast for a distance of thirty nautical miles. Canada, fearing it would be denied access to any of the ports, contended that its sovereignty extended westward to the crest of the mountains nearest the coast. When the United States uncharacteristically sent no troops or civil authorities to enforce its position, Canada dispatched the Mounties, including a sentry with a Maxim gun, to back up its own. The matter was not resolved until 1903, by which time the Klondike rush was over. The Canadian position was so weak in law that even Canada's greatest anti-American historian was grudgingly forced to concede the point based on the evidence.[5]

One function of the Mounties, who were sometimes charged with small-scale graft, was to interdict smugglers, particularly those who tried to bootleg whisky. Another was to keep out undesirables, which in practice meant members of the Smith organization who might try to penetrate the border in disguise. Their major function, however, was to prevent tragedy by refusing to admit anyone who did not have a year's supply of food. The welter of hurriedly published Klondike guidebooks, including a popular one from so distant an authority as the Chicago *Globe*, offered specimen lists. These might include four hundred pounds of flour, two hundred pounds of bacon, and one hundred each of beans and sugar, down to twenty-four pounds of coffee and fifteen of salt.

Building on recent developments in the military, the gold rush provoked technical advances in the canning and dehydration of food and greatly increased the variety available. But even with condensed and evaporated foods, a miner's supplies could weigh one ton. Near the foot of the pass was an area called the Scales, where the trekkers weighed what they had and repacked it, usually in units of fifty pounds each. For that was as much as the average European could carry up the slope. At that rate, a person had to make at least fifty return trips, taking many weeks in all. The base of the pass was an eerie streetscape of boxes and crates that bore miners' names and were slowly buried in deep snow as the owners trudged back and forth, like birds building nests.

At this point, the Native people began to profit from the Klondike gold rush in a way that indigenous people had never done during previous rushes. The local Chilkoot and two other Tlingit nations, the Chilkat and Stikine, hired themselves out as professional packers to the whites, whom they despised. They charged a goodly rate, which increased as the rush intensified. Following an incident in which an American paid a packer in Confederate States bank notes, they accepted only gold and silver, and they were sometimes known to demand a pay rise halfway up the pass or to dump the goods they were carrying if a better offer came their way. The whites must often have been infuriated, but they could not match the strength of the Native packers, most of whom routinely carried one hundred pounds, and a few of whom performed extraordinary feats. One packer carried three hundred pounds; another is said to have hauled up a hogshead weighing three hundred and fifty pounds. Jack London claimed that he beat Natives in races to the top.

White entrepreneurs were also drawn to this final leg of the upward journey, where the grade was about thirty-five degrees. Various people hacked steps out of the ice to make the going easier and collect tolls, but it could still take six hours to climb the last three hundred yards. Soon there were five tramways carrying goods overhead. The simplest was a rope on two pulleys powered by a horse at the top (one of the very few horses to reach that point). The most complex was a steam-powered system of freight cars shuttling eighteen hundred feet in the air on a copper cable strung from concrete pylons set at intervals in the mountain face. It was the work of the slightly misnamed Chilkoot Railroad and Transportation Company, whose president and promoter was later the American ambassador to France.

There were tragedies, when it seemed that natural forces were reasserting themselves in retaliation. In September 1897 the glacier above released a huge barrage of water that ruined the outfits of forty people. Among the victims was Arizona Charlie Meadows, a one-time frontier scout (he claimed to have fought Geronimo *mano a mano*) and Wild West star; he had been making the climb with a portable bar and casino that were to be his livelihood. In April 1898 an avalanche took sixty or more lives, despite the best efforts of a thousand rescue workers digging through snow as much as thirty feet deep. Soapy Smith appeared in the role of coroner and relieved the corpses of jewellery and money as they were deposited in a makeshift morgue. One of the few survivors was an ox capable of hauling six hundred pounds of gear; for reasons that now can only be guessed, it was named Mark Hanna, after the Republican Party kingmaker.

The Chilkoot Pass was a democratic funnel through which all, regardless of their station or ingenuity, had to pass, hand over hand, one behind the other. Once the summit was achieved and the formalities had been conducted, the flow grew wider once again. On the other side lay Lake Lindeman, connected by a canyon with Lake Bennett, into which the White Pass trail now emptied, so that the two streams of humanity merged into one stream with a common task. From this point, the way to Dawson was by water, through the lakes and then the Yukon River, a journey presenting an impressive number of natural obstacles, some of them as daunting as anything the miners had experienced so far. Marsh Lake, Miles Canyon, Squaw Rapids, White Horse Rapids, Lake Laberge, Five Finger Rapids, Rink Rapids, Split-Up Island, and Split-Up City. Some of the names alone were enough to give one pause.

The first task was to build a boat. At the height of the rush, thirty thousand people turned a sixty-mile stretch of waterway into a giant boatyard. The inevitable tent communities, including casinos and chapels, sprouted wherever enough people paused to carry out their rough labour. Turning standing timber into planking was gruelling work. The sawpits (actually scaffolds) witnessed the final break-up of many friendships that had somehow endured every other hardship and frustration along the way. Pierre Berton tells the story of two childhood friends who had married sisters and had been partners for life, but who could not restrain themselves when building a boat and could not repair the damage done once they lost their tempers.

Few of the builders can have had experience at the trade, but many had ingenious designs in mind. As a result, the flotilla that presented itself to the river in June was one that would probably never be rivalled for variety, not even by the fleet that evacuated Dunkirk early in the Second World War. There were homemade canoes and store-bought canoes and scows weighing twenty tons. There were hollowed-out logs, outriggers, kayaks, rafts in unbelievable variety, even a miniature sidewheeler in the Mississippi style – in all, 7,124 different craft carrying an estimated 30 million pounds of food alone. The first weapon in the gauntlet was Miles Canyon, whose steep black walls were at one point only thirty feet from each other, thirty feet separated by a whirlpool in which, it is recorded, two Swedes in 1895 had been trapped in a fast-swirling, ducking motion for six hours. In a matter of days, one hundred and fifty boats were wrecked, with the loss of five lives, in the first five miles of the downriver journey, before even reaching the White Horse Rapids. The Mounties intervened, forcing all women and children to walk the portage. Under threat of a hundred-dollar fine, the males had to stop until their boats were seen to be strong enough and in capable enough hands. Thereafter, there were far fewer wrecks and only eighteen more accidental drownings.

At Lake Laberge, there was another Mounted Police inspection post, beyond which lay Five Finger and Rink rapids, a short distance beyond the hilltop home of George Carmack, who must have wondered at what he had wrought as the steady flow of queer boats moved past him without apparent end. At the mouth of the Pelly was Split-Up Island, a place so named because now, with the end in sight, people who had survived all the foregoing hardships lost their last reserve of fellowship and fell to fighting. The curious phenomenon was so general that the Mounties once again had to intervene. Split-Up City, farther on, had a similar effect on people, for it lay amid a confounding network of islands and back channels where many travellers, after having come so far and survived so much, discovered they were lost.

There were any number of alternative routes to Dawson, none of them totally satisfactory, some of them a virtual guarantee of failure, all of them in use to a greater or lesser degree before the main stampede got under way in 1898. In the autumn of 1897, for instance, eighteen hundred people attempted to go by what was supposedly the easiest way, an all-water approach nicknamed the rich man's route. On a map it seemed straightforward enough: steamer passage from Seattle to St Michael, then steamboat connections up the Yukon River, for a total of 4,700 miles. But only forty-three of the eighteen hundred made it to Dawson that year, and of those only eight had enough supplies to remain.

One of the worst experiences was that of a group led by W.D. Wood, the mayor of Seattle, whose party spent three weeks building its own riverboat on the shore of Norton Sound. Another expedition was there for the same purpose, and the two vessels were entrapped by ice eight hundred miles short of their destination, where a temporary town called Suckerville sprang up. It was complete with the traditional miners' meeting to dispense rough justice, which in this case meant appropriating the supplies with which Wood had hoped to establish a shop in Dawson, where his party did finally arrive 314 days after setting out. Had the two boats made it only another hundred miles, they would have been subsumed into Rampart City, a community of one thousand built around the discovery of gold in 1896 by one of the mixed-race Russo-Natives who had stayed in Alaska after the country had been sold to the Americans. One of the people they would have met would have been Rex Beach, the future novelist, who was stuck there while making his way to Dawson.

American cities such as Seattle and Canadian ones such as Vancouver and Victoria were in fierce competition for the market in Klondike goods, and they helped fuel the notion that one should if possible travel to the goldfields by either an all-American or an all-Canadian route, as the case might be. The popular imagination did not grasp enough geography to know that there was no such thing as an all-American route. Patriotism coupled with a desire to avoid Canadian customs triumphed over reality. Similar motives drove Canadians and Britons to follow some of the all-Canadian routes, which could be pursued only at the most horrible cost to animals and humans alike.

The so-called all-American route meant an assault from the south, landing at one of the inlets on the Gulf of Alaska and trudging overland – over ice, to be accurate – to the Tanana River, which flows into the Yukon River. This might mean landing at Cook Inlet on the west or at Yakutat Bay, hundreds of miles farther east at the corner of the panhandle. But the majority found themselves going up Prince William Sound to the port of Valdez, a scurvy-plagued agglomeration of huts and tents to which they were drawn, in many cases, by an erroneous report that an old Russian trail led from there. In fact, what they found was the majestic but intimidating Valdez Glacier. They faced a twenty-mile climb to the summit (where a Christian agency soon erected a relief station), followed by a nine-mile slide down the other side. Much of the journey required steel claws on one's shoes to grab the ice, and travel was often restricted to the night, for during the day the crust was not always strong enough to support human weight. Snow blindness was another hazard of daylight travel.

Hopeful climbers began arriving in the autumn of 1897, thanks to quick action by one of the American steamship lines. Many brought animals that became ill or injured on the journey and had to be shot and thrown overboard. It was February 1898, however, before the first of the thirty-five hundred passengers who opted for the route could get under way, owing to the deep snows. A few people died in an avalanche. Partnerships broke down. And on the other side lay a twenty-five-mile stretch of rapids in the Kluteena River, which only two hundred people managed to get through that year. Conditions on the other all-American alternatives, the Cook Inlet and Yakutat Bay routes, were even worse. Yakutat led to the Malaspina Glacier. In one group of a hundred people, many were crippled or blinded permanently and only forty-one ever arrived at Dawson.

The all-Canadian routes were more numerous and just as arduous in their different ways but on balance only slightly less deadly. The Ashcroft Trail led one thousand miles from Ashcroft, northeast of Vancouver, through the worked-out diggings of the Fraser and Cariboo rushes and across the Stikine River until it reached Teslin Lake, where the Yukon River begins. On the way it passed through Telegraph Creek, so named because in 1865 the Western Union Telegraph Company had begun to push a line through the country in an attempt to link North America and Siberia; the scheme had been abandoned after a telegraph cable had been laid under the Atlantic, but the poles were still visible. It is estimated that fifteen hundred people (the novelist Hamlin Garland among them) and three thousand horses started out on this trail. The gold seekers gave names such as Poison Mountain and Starvation Camp to places along the way, and some left pitiful, despairing graffiti on trees and rocks; a number committed suicide. As for the horses, all of them succumbed, mostly because of the absence of grass on the northern half of the trail.

A variant course, a nearly all-Canadian one, began at Wrangell in the Alaska panhandle – a suddenly revitalized town from the Cassiar gold rush where Soapy Smith operated a sort of branch plant – and proceeded up the Stikine until meeting the Ashcroft Trail at Glenora, a little hole which by early spring 1898 had five thousand people, including the Reverend John Pringle, a Canadian Presbyterian missionary who later became one of the Klondike's best-loved figures. There was talk of a railway to follow part of the Stikine Trail, but although a dozen miles of right-of-way were cleared, political difficulties proved insurmountable.

The mercantile classes of Vancouver and Victoria promoted the Ashcroft route as vigorously and deceptively as their Seattle counterparts did the routes over the Alaskan glaciers. The shopkeepers of Edmonton – a town of only about twelve hundred clustered around the former Hudson's Bay Company fort on the North Saskatchewan River – showed considerable enterprise and even less sense of responsibility in ballyhooing the various Edmonton trails, which, even compared with the others, were not trails at all but vague scribbles. Among those who fell for the promises of a back door to the Yukon were Steamboat Wilson, the former mayor of Kalgoorlie, and Viscount Avonmore, an Irish lord who arrived with a large party, including servants.

One set of suggestions caused people to attempt Dawson by going northwestwardly along the Liard or the Pelly Rivers, a trip of about fifteen hundred miles. The assumption that water routes are by their nature easier led others to travel twenty-five hundred miles, taking the Athabasca River and Great Slave Lake to the Mackenzie River, then making their way west. The former was bad enough, the latter horrible; one-third of the nearly eight hundred men and women who chose it either turned back or actually perished, and no horses survived. The best hope from Edmonton was the Peace River route, another umbrella term. One could choose between a 320-mile journey, mostly by water, to Peace River Crossing, or a much longer overland crossing direct to another Peace River settlement, Fort St John; from either place, one selected from a menu of options for getting to the Pelly and thence to the Yukon.

There were two other important possibilities. Almost nine hundred people left Edmonton for Dawson taking the Mackenzie Strait towards the Arctic Circle. They added as much as one thousand miles to the voyage, but three-quarters of them made it, though the trip took some souls eighteen months. One couple chose this as their honeymoon trip. In another instance, a child conceived along the way was born before his parents arrived in Dawson. Athabasca Landing, above Edmonton, was soon full of boat yards catering to the Klondike bound, whose first obstacle was more than one hundred miles of rapids. The dominant nationalities can be deduced from the fact that the tent community had one set of streets with names such as Piccadilly and the Strand and another with names such as Fifth Avenue and the Bowery. A similar dichotomy existed towards the opposite end of the trail, above the Arctic Circle, at a place called Destruction City, which was given this name because an even worse set of rapids there wrecked many boats or forced gold seekers to cut down their craft to a size they could portage. This desolate little camp, where people died of scurvy, was dotted with symbols made of calico and flour sacks: the American red, white, and blue, and the red ensign of Canada.

8

Climax and Retreat

In 1897, while so many people were preparing to make the journey to Dawson, most of those who were already there were desperate to quit the place. They were escaping overland and by river, under even greater hardship than they had experienced coming in. They were fleeing starvation, a fate that nearly ended Dawson's effective life before its heyday could begin.

There were early signs that autumn that all was not well. The caribou herds were moving out early in search of better grazing, and Joaquin Miller (1839–1913), the western poet who was famous as a romanticizer of the California gold rush he was too young to have known, received a rude shock when he stepped off the wharf. Miller (born Cincinnatus Heine Miller) had come as a journalist, hired by Hearst to send back colourful descriptions of the goldfields. Like the other two hundred or so people arriving from the Outside at the time, he had made a point of travelling fast and light. He disembarked, he said, carrying a flute and a raw onion, and someone almost immediately offered him five dollars for the onion. The significance of the bid escaped him, however, and he reported to American readers that there would be no famine in Dawson. Inspector Constantine of the Mounted Police was of the opposite view.

Two steamboats of supplies were en route to the stricken town, now swollen with an influx of miners from the outlying creeks, but were trapped hundreds of miles away as the water level fell. With winter drawing near, the stores of the two trading companies were forced to ration what stock they had. The public eating places closed one by one. People began to leave for Fort Yukon, three hundred and fifty miles downriver, hoping there was more food to be had there.

A San Francisco man, Thomas McGee, led a party of fifteen on a small steamer towards Fort Selkirk, intending to abandon it there and dash overland to the Lynn Canal. The steamer made only thirty-five miles the first week and then broke down completely. Its passengers were forced to employ canoes. They reached Skagway after forty days and quickly boarded a ship that was being vacated by gold-frenzied newcomers full of hope. When the *Portus B. Weare*, one of the vessels that had first brought news of the goldfields to the rest of the world, put in at Circle City, a boarding party of heavily armed miners bought thirty tons of cargo at gunpoint. Two similar incidents followed. In one of them, at Fortymile, the hungry vigilantes kidnapped a young woman as well. When the *Weare* and another vessel finally arrived in Dawson, where men were reduced to ice-fishing and trying to hunt hare, it was learned that they carried no food, only liquor and hardware.

Constantine knew then that Dawson would have to be evacuated and that there was a slight chance that those who agreed to go could be moved to Fort Yukon (not that conditions there would be much better). The two vessels pushed off into a river choked with ice. The one made it to Fortymile only with the greatest difficulty before it was frozen fast. Then a chinook thawed the channel long enough for it to proceed a bit farther. It managed to reach Circle City but not Fort Yukon. The other ship, the *Weare*, almost erupted in mutiny at one point but managed to achieve its destination after one hundred men with axes hacked it free of its bonds. Supplies at Circle City and Fort Yukon did indeed turn out to be less than was hoped for, and there were even more incidents of violence or threatened violence. But even as the *Weare's* passengers were chopping their way through the ice, other people were forcing a river captain to take them in the other direction, towards Dawson. In all, perhaps nine hundred people made a successful retreat from Dawson that autumn and winter, some of them trudging between the pressure ridges on the solidly frozen river in temperatures as low as fifty degrees below zero Fahrenheit. Some who set down were forced to return, including Joaquin Miller, who lost an ear, a finger, and part of a toe in the process and was snow-blinded temporarily.

Those who spent the winter of 1897-8 in Dawson, voluntarily or otherwise, did much to illustrate the town's character and perhaps even to shape it. Inflation was staggering even by gold rush standards. A pound of rotten potatoes brought a dollar, a like amount of rancid flour three times as much. Speculators bought in quantity and even began hoarding whisky glasses to drive up the prices. Yet the cost did not actually matter, since the food was so precious that it was virtually priceless and gold so plentiful that it was comparatively worthless. The only people who literally starved to death were the Natives, especially the women and children out in even more remote areas, such as the country north of Fort Yukon. But it was on the whites in Dawson, not on the Natives, that outside concern focused. Businessmen in Seattle and other gold rush supply centres persuaded the U.S. government to send relief to the trapped miners. It came in the form of more than five hundred reindeer brought from Norway via New York and Seattle, along with nearly seventy Laplander, Finnish, and Norwegian herdsmen. A year later, 20 per cent of the animals straggled into Dawson, after weather and marauding wolves and hungry Native marksmen had exacted their toll.

The spectre of starvation did not extinguish the urge to wheel and deal, for entrepreneurial ability was an important survival skill. At the height of the Johannesburg gold rush there had been an incident in which someone offered a woman's sexual favours at public auction. In Dawson that winter, a woman climbed atop the bar in a saloon and put herself on the block as partner, housekeeper, and companion until the spring, with a third party holding the money in surety. She was a Québécoise named Mable LaRose, and she realized $4,000. Some powerful new personalities emerged in the business community. Belinda Mulroney had arrived in town with a load of dry goods and hot-water bottles and had sold them at a large profit with which she opened a restaurant and entered the property-development field. Now she went into partnership with Big Alex McDonald to salvage a shipment of food lost in a shipwreck on the river. She was in a strong position actually to prosper during the siege. And it is typical of gold rushes that the prospecting did not cease during this life-or-death period but seemed to intensify. There were several miniature rushes composed of the stranded Dawsonites, who on nothing more than rumour scrambled to reach creek beds as far as fifty miles distant. It was during this period, in fact, that a new category of wealth was generated when a German, William (Cariboo Billy) Dieterling, reasoning as Australians had done that the greatest concentrations must lie undisturbed in ancient stream-beds, staked the first bench claim on the Klondike. He was one of Dawson's rich men, but only one of many, when spring finally came, along with the thousands of boats, and Dawson became the most populous Canadian city west of Winnipeg, in fact the biggest in the whole Pacific Northwest.

Dawson's location was in every way inferior. Before the summer of

1898, the town had already undergone a serious flood and the first of the several fires that devastated it from time to time, ensuring that no buildings dating from before 1900 still stand there today. But it was a miraculous place all the same. It was far more remote and more conspicuously wealthy than any other gold rush community had been. The first of these facts bespoke a need for services, the second a demand for luxuries. As a result, Dawson blossomed overnight as a full-blown city. Its heyday lasted only a year, from the summer of 1898 to the summer of 1899, but in that period it had running water, electricity, steam heat, telephones, and motion pictures, some of which amenities were not found until much later in far larger centres. Dawson also had new Paris fashions and four-star restaurants, and an economic and social life that at times seems almost too colourful for what was after all so well policed and strictly administered a place.

From the beginning everything was up and down. The first rush of supplies following the dreadful winter brought eggs from Seattle at $1.50 each. Within a week there were so many that the price dropped to twenty-five cents, or the cost of a decent three-course meal in any other major North American city. And then someone who was truly clever imported a shipload of hens. When someone else arrived with tinned milk, another appeared with a cow. A person with a keener imagination brought in a boatload of kittens, which lonely prospectors hurried to buy. It seemed that almost anyone could make enormous profits by importing either the most mundane goods – butter, wellingtons – or else outrageous luxuries such as ostrich feathers or the hats to stick them on.

Curiously, one of the most reliable commodities was news. The world was eager to learn about the Klondike, and the major American, Canadian, and British newspapers had correspondents on the scene. These included Elizabeth Cochrane Seaman (1867–1992), the famous Nellie Bly of the New York *World*, whose daredevil style of journalism had led to books such as *Around the World in Seventy-two Days*. But the Klondikers were still more eager, frenzied almost, for news of the Outside, what with the Spanish-American War (from April 1898) and the Boer War (October 1899). A gambler made arrangements to import bundles of obsolete San Francisco newspapers and always sold as many as he could get at ten or fifteen times the published price even though they were out of date. Thus, when a comparatively fresh copy of the Seattle *Post-Intelligencer* materialized, the result was an auction at which it was sold for fifty dollars to a miner who then hired a local lawyer to give public readings from it. The miner had outbid both of Dawson's dailies, the *Midnight Sun* and the viciously pro-American *Klondike Nugget*, edited by Eugene Allen, who held the honour of beginning publication first. One of the *Nugget*'s best-loved assets was a sixty-five-year-old news-vendor named Uncle Andy Young, who often sold a thousand copies of a day's paper. His was a profession in which large gratuities could be counted on, though Charley Anderson (the Lucky Swede) once pushed inflation to absurd heights when he gave a newsboy fifty-nine dollars in gold for a single paper.

After operating a Dawson branch for only two weeks, the Canadian Bank of Commerce shipped out $750,000 in gold dust. This was the bank for which Robert W. Service came north to work long after the rush had ended. Like its rival in town, the Bank of British North America, it accepted commercial-grade dust, mixed with black sand, at eleven dollars an ounce, paying the full sixteen dollars only for the higher-grade dust, which fraudulent operators sometimes augmented with brass filings. Arbitrage therefore became a profitable occupation, as did the weighing of gold, for which merchants, saloon owners, and casino owners paid twenty dollars a day. Gold was so awkward, especially for large transactions, that paper money, almost any paper money, was worth more than its face value. Today, the Bank of Commerce notes issued with the legend DAWSON or YUKON are collectors' rarities; even then, they were prized for their utility. The notes of 'broken banks,' which had proliferated in the United States before the Civil War, passed current in Dawson, though many of them had been worthless even when they were new. So did Confederate money of the sort refused by Native packers on the Chilkoot Pass. The same applied to the currencies of most of the nations represented in the gold rush, which included Afrikaners, who now had practical gold-mining experience of a high order, and Japanese, who had never before taken part in a gold rush.

To the modern observer, one especially curious aspect of Dawson social life was the almost comic dependence on nicknames. One might ascribe it to the strange devotion to secrecy that has always existed in the United States alongside an equally strange love of unsolicited candour. Yet the phenomenon is not found to nearly the same extent in the records of the California or Colorado rushes or in those of the Comstock Lode. The spread of nicknames in Yukon seems to have been a reaction against the rising level of education among the stampeders and against the general pressures of civilized behaviour. It was related to the type of people who were attracted to the Klondike. Thornton John, the elder brother of the English painters Gwen and Augustus John, heard the news and 'left for Canada, taking no baggage, only coming back to enlist in the First World War, after which he returned to the gold fields.'[1] There are numerous recorded instances of Klondike miners who never travelled without their works of philosophy or sets of the classics of English literature, and they are probably more than a reflection of the fact that Canadians in the extreme North are traditionally voracious readers, particularly during the long winters. It speaks to the fact that by 1898 the world was closing in and that different types of people, for different reasons, wished to outrun the tide. As in a small primitive society, such as the one they were creating in their imagination, people's names often derived from their profession. Faro dealers were by custom called Kid, whereas Waterfront Brown was a debt collector who patrolled the steamboat docks searching for attempted escapers. Or a nickname might stress one's professional experience, as was the case with Cassiar Jim and many others. Or it might refer to a person's appearance; both Spare-rib Jimmy and the Evaporated Kid were exceptionally slight of build, while Spanish Dolores was probably distinguished by her Latin complexion.

To be sure, there were people who used the Klondike as a hiding place. The story has come down of a private detective from Chicago, C.C. Perrin, and his obsessive hunt for an escaped murderer named Frank Novak. Perrin spent six months peering into the faces of men coming through the Chilkoot Pass until he finally got his man. Then there were the veterans of previous frontiers who now, with the century and their lives both drawing to a close, saw the gold rush as a means of disarming the future by reliving the past. The best known of these was Martha Jane Canary (1848?–1903), the famous Calamity Jane. There were others, however, who were almost as well known in their day, such as Captain Jack Crawford, 'the poet scout,' who operated a general store where Dawsonites could purchase ice cream.

In the United States, foreign nationals were no longer permitted to hold mining claims. Canada was in no position to introduce similar strictures because without foreign capital and manpower the North-West Territories, which then included what later became Saskatchewan, Alberta, and the Yukon Territory, could not be developed. There was worry nonetheless that the open-border policy could backfire. The American pattern of frontier development included many instances in which American settlers, prospectors, and speculators poured into an area until they far outnumbered the indigenous inhabitants and then in effect established their own independent governments before offering themselves for inclusion in the federal union. It was by this means that Texas and California had become states – not to mention Washington State, in the former Oregon Territory, which might otherwise have become a Canadian province. Clearly, there was the same threat in the Klondike, where Americans outnumbered others four to one and where the border between the two countries was already in dispute. Accordingly, the Yukon district was separated from the North-West Territories to make it easier to administer and to reinforce Canadian claims to sovereignty. More than two hundred Canadian troops, approximately 20 per cent of the standing army, were designated the Yukon Field Force and sent north via Telegraph Creek to assist the Mounties in keeping order and to protect Canadian interests. They carried two pieces of artillery, which now repose outside the RCMP station in Dawson, and two Maxim guns. They were accompanied by four members of the newly organized Victorian Order of Nurses and two newspaper correspondents. (One of the latter was Faith Fenton [died 1936] of the Toronto *Globe* whom the field force's commander accused of inciting disorder among the other women. She seems to have fitted in fine in Dawson, where women could be enslaved, almost literally, but where some could use the prevailing spirit of individualism to advance themselves in ways not always possible in the south. Dawson had a female physician, for example, long before most towns on the Outside.)

That order was indeed maintained with no sacrifice of economic benefit can be attributed largely to Sam Steele and his colleagues who knew on which points to stand firm and on which to compromise. The prohibition against firearms was enforced rigidly. A great many of the Americans, perhaps a majority, had tried to enter the territory carrying pistols, and some must have succeeded, to judge by the few accounts of suicides and other occurrences involving side arms. Shoulder arms were confiscated as well, and these were purchased in bulk lots by miners who used the barrels as steam pipes for thawing frozen ground. As for gambling, the Yukon was, as it remained for many decades, the only place in Canada with legal casino gambling. Moreover, the gambling in Dawson was unrestrained, in the grand manner of the American mining towns, though flagrant cheats were punished when caught. The most famous casino was the Monte Carlo, but there were others nearly as well known, such as the Bank Saloon where Front Street, the principal thoroughfare, met King. It was operated by the city's best-known gambler, Silent Sam Bonnifield (1866–1943), who had worked his way there after taking part in various American rushes.

The casinos were also dance halls. The high-kicking and usually pseudonymous dancers more nearly resembled courtesans than common prostitutes; the most storied was Gertrude Lovejoy, known as Diamond-Tooth Gertie because of her singular dental work. Their performances were considered daring, though Sam Steele intervened when Freda Maloof, a Greek immigrant who was billed as 'the Turkish Whirlwind Danseuse,' announced her intention of recreating the striptease with which Little Egypt had scandalized the Chicago World's Fair of 1892–3. The Mounties also enforced the Lord's Day laws and even made attempts to uphold the ordinance against improper language. Punishments were harsh: fines, hard labour at the government wood pile, expulsion from the Klondike, or some combination of all three.

Although Dawson had its obligatory opera house, as any purpose-built theatre on the frontier was called, the dance halls did much to fulfil the demand for stage entertainment. They witnessed productions of *Camille* as well as boxing matches, though the latter were certainly more common, given that Dawson's appreciation of the ring rivalled Johannesburg's. The British Empire heavyweight champion was Frank (the Sydney Slasher) Slavin, an Australian. He had been touring the United States, putting on exhibitions for audiences of appreciative yokels, but moved north with his sparring partner, Joseph Whiteside (Joe) Boyle (1867–1923), who soon became an important figure in Yukon mining. Alexander Pantages, whose name later graced one of America's two largest theatre chains, began his career in Dawson, unlike Sid Grauman, the future proprietor of Hollywood theatres, who did not enter the field until after leaving the Yukon.

As in a number of American mining towns, the red-light district was called Paradise Alley, but it was not totally laissez-faire. Sam Steele had at first banned the prostitutes from appearing in public until 4 PM each day. Only when that failed to bring some order did he relegate them to a small area. It consisted of perhaps seventy cribs, or log cabins in this case, with the name of each occupant above the door. The wretched creatures – usually recent immigrants, frequently Belgian, or so it was claimed – were cruelly manipulated by their maques, or pimps, and were later removed to another area called Hell's Half Acre. The men who patronized them rationalized their behaviour by an elaborate code of chivalry towards other women, even if their status was only a little higher. Pierre Berton relates the story of a coroner's inquest into the suicide of a dance-hall girl at which six different men who had been sharing her bed testified to her purity.

The give and take between authority and licence therefore operated on a number of levels, both public and private, in doses ranging from unstated tension to open confrontation. At the back of it all was the difference between the United States and Canada, which came to a head in disputes over how the production of gold should be regulated. The question is complex. On the one hand, the Americans did tend to ride roughshod over the political environment, insisting that there was and should be no distinction between Dawson and that which existed only fifty miles away across the international boundary. Americanism was a force in the community and the post of American consul an important position (though the person who occupied it for most of the gold rush was such a buffoon that he had to be dismissed, despite the special providence that protected buffoons who contributed heavily to the Republican Party). Their insensitivity was matched only by the speed with which they took offence. They were determined that 4 July should be a public holiday, with fireworks and parades in the streets of Dawson, whose names tended towards Wall Street and Broadway. They grew testy that anyone should resent such self-love. Sam Steele showed finesse in proclaiming that since Dominion Day and 4 July fell but seventy-two hours apart, it would be efficient to celebrate them simultaneously. On the other hand, the Americans, and others too, seemed justified in their complaints about the incompetence and dishonesty of the Canadian administration. One reason that newspapers, local and domestic, were of such importance in the Klondike was that the postal service was abysmal. Letters could take many months to reach Dawson when not lost entirely, and their arrival was no guarantee of delivery. Stamps were tightly rationed and the queues so long and persistent that a number of people, usually women, who benefited from the miners' chivalrous code of conduct, earned their living as proxies.

A more serious cause of grumbling was the government's tax on gold, introduced as early as September 1897. It was the first such tax ever imposed during a gold rush, but then this was the first gold rush in which most of the product was shipped outside the country. At first, the tax was a straight 10 per cent royalty on all gold mined, but this was later boosted to 20 per cent in the case of mines producing more than $500 a week. By the time the rush was at its crest, the commissioner of the Yukon, Major J.M. Walsh, the famous Mountie who had laid down the law to Sitting Bull following the Custer massacre, brought back the straight 10 per cent royalty and added an annual $5,000 exemption. None of the amendments did much to quiet the protests, official and otherwise, or to stop the enormous traffic in unreported gold which was the natural result of these government edicts.

Another source of discord was graft in the office of the recorder, where claims and ownership changes were filed. Here, too, people spent days in queues. Women once again were given preferential treatment, though this had the effect of benefiting men, including sometimes even the clerks themselves, who hired prostitutes and dance-hall girls to file for them. In any event, the ledgers were a shambles of inaccurate, altered, and even forged data. Not only was a small bribe necessary to have a claim recorded, but it was enough to have a prior claim misplaced or invalidated. The government abolished fractional claims, but these still seemed to be awarded to a few favoured individuals. Surveyors grew prosperous, and so especially did litigation lawyers. The last straw was the lifting of a seven-month ban on new activity in the Dominion Creek district, forty miles away. A small stampede broke out. Then came to light substantial evidence that Commissioner Walsh or his friends and colleagues had profited from advance knowledge. Walsh was replaced by William Ogilvie, who convened a royal commission, though by then the trail had grown cold and many of the principals had moved on.

As the tone of these developments suggests, by 1899 the Klondike gold rush was rapidly losing its youthful appeal. The easy alluvial gold was becoming scarce, and capital was required to retrieve what remained. Gold dredges were used to tear up the creeks. The early ones were of the continuous-bucket type that had been perfected in California and New Zealand at a similar stage in the development of the goldfields. Soon Joe Boyle had three of the largest dredges in the world working in the Klondike. A decade later, during the First World War, he raised a Yukon detachment. Later still, he became the most credible of those who claimed to have been the lover of Queen Marie of Romania (others included the journalist Gene Fowler and the playwright Charles MacArthur). More importantly, he became the Barney Barnato to Arthur Treadgold's Cecil Rhodes.

Treadgold (1863–1951), a descendant of Sir Isaac Newton, had abandoned a career as a classicist at Oxford in order to hurry to the Klondike as correspondent of the *Manchester Guardian*. Once the small fry withdrew, Dawson, following the pattern of previous gold rush towns, grew insular and then withered (though it remained the territorial capital until 1953). Amidst this economic shrinkage, Treadgold was carrying out his fifty-year struggle to consolidate the extraction firms under his own Yukon Consolidated Gold Corporation, an achievement he finally realized, however briefly. In the final phases of the gold rush, reputations began to exceed accomplishments as the room for individual initiative became smaller and smaller. M.A. Mahoney, who later discouraged the story that he had carried a piano over the Chilkoot Pass but did not object to the names Klondike Mike and King of the Klondike, provides a clear example of this phenomenon. It is significant that although 'sourdough,' a homely term for a lowly type of individual, was in use as early as 1864, the distinction 'gold king' dates only from 1898 in Canada and became current only as the era of the small prospector was drawing to a close.[2]

Outside events were also an important consideration, though their exact effect is difficult to measure. The death of the old Queen in 1901 was some sort of harbinger of a new age, and the Americans' war on Spain was a fresh channel for their destructive enthusiasm. Major Frederick Russell Burnham, DSO (1861–1920?), an American who went to Africa to be a prospector but wound up as a mercenary working for Cecil Rhodes, came back to North America, heading first for Yukon and then Alaska. The fact that he was in the bush and missed his chance to fight alongside Theodore Roosevelt was to rankle with him for the rest of his days. Later his luck improved. He was in Skagway when a steamer put in carrying a message asking him to serve the British against the Boers. He was packed and aboard when the same vessel departed two and a half hours later. 'Although Mr. Burnham has lived in Skagway since last August, and has been North for many months,' a local newspaper reported, 'he has said little of his past, and few have known that he is the man famous over the world as "the American scout" of the Matabele wars.'[3]

No doubt the real sign that the Klondike rush was over came at midsummer 1899 when two or three thousand Dawson residents, having stampeded in, promptly stampeded out, on hearing news that another gold rush was under way at Nome, Alaska. Gold had been discovered there a full year earlier by a Swedish missionary and four others, but this time it was Dawson's turn to get the news late. Nome's location on the Bering Sea, in relatively easy communication with the Outside as well as with St Michael and other Alaskan centres, gave the advantage to those well to the west of the Coast Range. No more than half a dozen whites had lived among the Inuit who camped there during the seal hunt. By the winter of 1898–9, forty men had proclaimed a mining district and had staked seven thousand acres of placer claims, and it was not long before a government was established. It was a corrupt government even by the standards of the American frontier. There was a sense of desperation and frenzy in the greed, which contrasted sharply with the more leisurely if better organized machinations of Soapy Smith. Thus, Nome provided an even stronger contrast to Dawson than Skagway had done. People seemed to sense that Nome was their last chance.

American mining law had never quite been synchronized with the cycle of booms and rushes. The first comprehensive federal mining statute was passed by Congress in 1866, when corporatism was growing and individualism was in retreat. It was replaced in 1872 by a new act that failed to correct the imbalance and indeed favoured large organizations even more. There were contradictory views on whether foreigners could locate or work claims. Loopholes that permitted companies to use jitneys to retain claims of virtually unlimited size were not closed until 1912. Significantly, that was the year when the Arizona Territory, the last part of the U.S. mainland to be developed by gold or silver rushes, became the last of forty-eight contiguous states. There was thus a framework for legal as well as social chaos when in July 1899 shallow alluvial gold – ruby gold, some called it – was discovered on a beach at Nome that ran hundreds of yards until it merged into tundra.

A government geological bulletin stated that there were two thousand miners on the site averaging twenty dollars a day. A Seattle newspaper spread a report of some people clearing a thousand dollars a day. The great British editor W.L. Stead made arrangements for a correspondent, who even before arriving at the scene wrote: 'Many believe this inhospitable shore to be the richest placer-ground in the world. Miners, who came out of the district for machinery or food-supplies before last Winter double-barred the door – said that it was a case up there, not of gold mixed with sand but of sand and gravel mixed with gold. "If you walked along the beach, you could wash gold from your boots when you went back to your hut."'

The implication that solitary prospectors could still get rich summoned individuals of many nationalities, especially Americans. The detective who had been in charge of the Lizzie Borden axe-murder case was one. Another was a woman who brought her family's silverware for use in the restaurant she hoped to open. They flooded in by a variety of routes. As had happened so recently following news of Carmack's discovery, steamship companies, many of them fraudulent, began to disgorge passengers from almost any old hulk that was capable of making the journey. There was no harbour, only the long curve of the ocean shoreline, and launches ferried people and equipment to the beach. As at Skagway, it was a race to get one's goods to safety before the salt water destroyed them. But the confusion was more general than at Skagway because the only possible town site for Nome was in the middle of the area being mined so feverishly, all of which had already been staked by the time the steamships arrived. There was a rash of claim jumping such as no previous gold rush had witnessed.

In Dawson, where there was timber, accommodation had improved as the population rose. In Nome this was not practical. In May 1899 there were fifteen hundred people living in tents and lean-tos. The most substantial structures were made of driftwood insulated with peat. In such habitations people spent the eight-month winter. Tools were similarly crude and makeshift; sluice boxes were banged together from old packing crates, and rockers were lined with copper sheathing from the hulls of ships or even with silver coins hammered into sheets. But conditions were only slightly better the following year, when the human tidal wave from Seattle and San Francisco finally struck and Nome reached a peak population of perhaps eighteen thousand.

One of Nome's most bizarre figures was Raymond Robins (1873–1955), who had thrown up his profitable law practice in San Francisco to go to the Klondike and was now, following a spiritual rebirth, both the religious leader of Nome and one of its eminent racketeers; he later achieved renown as a social economist. His sister was the actress and suffragette Elizabeth Robins (1862?-1952), famous for her interpretations of Ibsen in London's West End. She went to Nome to visit her brother and left a vivid account:

The beach is crowded with people and stores, already landed. Boats have been coming in pretty steadily since the end of May ... The space remaining is already piled with freight – food supplies; barrels of beer and whisky; bags of beans and flour higher than my head; lumber, acres of it, extending beyond the tents and up on the tundra; furniture, bedding, pots and pans, engines and boilers, Klondike Thawers [a type of stove?], centrifugal pumps, pipe and hose fittings, gold rockers, sides of bacon, blankets, smart portmanteaux and ancient seachests – as odd a conglomeration as ever eye rested on. We make our way through the mass, stepping over ropes that secure small boats, getting out of the way of the few teams that lord it over pedestrians and that yet go perforce so slowly, half the time the wheels on one side scraping against the mountains of provisions, the other side in the water, so narrow is the space between the high piled freight and the sea.[4]

As for the human beings to whom it all belonged, Robins discerned a minority who were 'bright and keen ... setting up their little tents and moving about between the tundra and the beach with elasticity and pleasure – the old story.' But most of the people seemed to her 'like so many Robinson Crusoes each on his own desert island. I never saw such loneliness in them, such outward showing of the inner sense of isolation.'[5]

There was an outbreak of smallpox, to which the local administration responded with pest-houses in the form of large yellow tents, while forcing ships suspected of carrying the disease to stand offshore for a quarantine period, flying the yellow flag. Nome also took the precaution of organizing a fire brigade, in contrast to Dawson, which had formed one only after the first devastating fire had run along Front Street. But such instances of civic betterment give a favourably distorted view of Nome's government, which consisted of a mayor and aldermen, a city attorney, and a police chief. The last of these was appointed by the first, following dishonest elections. An unusual element in the charter gave officials the right to impress fellow citizens into service during an emergency. There was thus a sort of standing vigilance committee that could be used to forestall any attempt to stop the profitable activities that ran unmolested. One observer wrote: 'Robbery of every description was in full swing ... At one time it was proved here that all three sheriffs and deputy sheriffs had done time in the penitentiary. One of them made his living by robbing the drunks he arrested. One day two men and I were given a dose of knock-out drops ... As we had had a drink in three different saloons, it was impossible to tell which one had done the trick. Two of us were already busted and lost nothing, but the other man was robbed of six hundred dollars.'[6]

The mayor, T.D. Cashel, was a booster, like mayors everywhere. 'Nowhere else in the world has there been discovered such a beach as this,' he said. 'Where the facilities for saving the gold have been at hand, the beach mines have outranked the richest placers of Gold Hill, Adams Hill, and other celebrated beaches in the Klondike.' But he was a mayor without civic authority. In 1900 there was a tax revolt when the administration tried to raise a levy for civic improvements, and only by the presence of U.S. soldiers was order maintained. Thus, an official such as the police chief, a man named Eddy, was not dependent on the public purse. He paid his men's salaries out of his own pocket, which he had lined with fines extracted from prostitutes and gamblers. The jail was privately owned, and the landlord threatened to terminate the lease and turn the prisoners out into the street. City officials converted City Hall to office space and kept the rents for themselves.

The federal government presence in Nome was even worse. Judge Arthur H. Noyes arrived in the town in the company of the man who in effect was employing him – one Alexander McKenzie, a corrupt political boss from North Dakota, who was lobbying Washington to expand and fortify the provisions making it illegal for non-citizens to dig for gold. He had Noyes appoint him to the post of receiver, which gave him effective control over a number of disputed mining claims and all the gold they produced. This was too blatant even for Nome, and petitions for redress were sent to the authorities in San Francisco. In 1901 Noyes was removed from the bench, part of a clean-up that also saw several city officials given short jail terms. As for McKenzie, he was, as the historian of gold rushes W.P. Morrell put it, 'taken off to San Francisco in judicial cold storage and punished with that salutary absence of severity which such an important political occasion demanded.' That is, President McKinley pardoned him after he had served ninety days of a one-year sentence.

On two occasions when the alluvial gold threatened to peter out, new beds were discovered farther up the beach, and the centre of activity moved nearer the bare hills that began four miles from the shore. Although buildings became more substantial, the place never achieved Dawson's level of civilization or its size. Having easier access and egress, people moved in and out more frequently, and only about twenty-five hundred souls spent the winter there each year. Many Klondike figures were attracted to Nome for its enterprise. Tex Rickard, for example, seems to have found it a promoter's paradise. But it was too rough and ghastly for others, such as Mike Mahoney, who moved on at the first word of a smaller gold strike elsewhere in the territory. The spiral was generally downward, however, as it was with all such gold rush towns; the difference was mainly one of degree.

Each year, as the pool of gold decreased and the percentage of beach gold decreased even more, mining moved farther afield and became more expensive. There was less opportunity for the prospectors and greater potential for the mining-company operators. The former availed themselves of every little rush of excitement that presented itself, as in July 1902 when an Italian, Felix Pedro, found gold on the Tanana in the interior, or the following year when news arrived of a Japanese discovery at what soon became the city of Fairbanks. Of those who remained, some found themselves drawn into the world of the businessman, borrowing money at 2 per cent a month for the steam-pointing equipment they now required. Others became wage slaves in the big mines and took part in the labour wars that erupted in 1907 and 1908. Still others became the Alaskan equivalent of tenant farmers, working other people's claims for a quarter or half of what they found.

When the Klondike rush ended, Dawson's population evaporated and its infrastructure imploded, but the place decayed with a certain grace. When times started to turn bad in Nome, the city shrank but retained its core of violence. Or perhaps it was that the violence based on easy money was replaced by that special violence bred of despair at the very absence of money. The Yukon remained a wilderness, but Alaska remained a frontier. That is the essential difference, and one the popular culture has often missed. There is a line about a 'shooting in the Klondike' in the song 'Put the Blame on Mame' that was made famous by Rita Hayworth in the film *Gilda* (1946). But it was of course in Alaska and not the Klondike that this particular kind of tragedy remained obvious even as the gold rush receded and finally disappeared. It was for killing a bartender in an argument over a prostitute in Juneau in 1908 that Robert Stroud (1890–1963), 'the Birdman of Alcatraz,' was first sentenced to prison.

Like the Klondike, Alaska attracted a number of figures from an earlier frontier. They sought a taste of an older America, but what they found was one quite different in tone. Wyatt Earp (1848–1929) was the manager and resident dealer at a Nome establishment called the Dexter Saloon, but he suffered the humiliation of being publicly slapped in the face by Bob Lowe, the town marshal. He seems to have tried other Alaskan towns before returning to his home in California and living out his retirement breeding racehorses and dealing in real estate.[7] There is reason to believe that Robert LeRoy Parker (born 1866), the notorious Butch Cassidy, did not die in Bolivia in 1908 along with the Sundance Kid but escaped and went to Alaska seeking gold and died in an old age home in Washington State in 1937. There is also some evidence to suggest that Earp met Cassidy while in Alaska and penetrated his false identity.[8] If the assertion is true, it might well explain Cassidy's departure. Even if it is not true, it speaks to the bittersweet mood of the Alaskan gold rush. The forces of time and change had just about caught up with one type of individual and one style of personal and political economy.

9

Last Stands

Insofar as ordinary individuals are concerned, the rushes in the Klon-dike and Alaska, with their hypnotic appeal to the dreams of people caught up in the drudgery of industrialized culture, represent the peak of the gold rush movement. Never again would the attraction be quite so powerful to such a variety of people or to so many. The gold rushes of the far Northwest also established the supremacy of the professional prospector, thus making the formula more complicated. When the next important series of events took place, in northern Ontario, there were four elements in the equation – remnants of the traditional amateur prospectors, the new breed of professional prospectors, government, and capital, the latter two of which had been gradually increasing their dominance in successive rushes. Each tried to realize some advantage; all worked against a backdrop of what by then was a familiar set of circumstances.

As in all the other venues to date, the presence of gold had long been known or at least suspected. As early as 1686, Native people in the region around Lake Timiskaming had led a Montreal captain, le Sieur de Troyes, to an outcrop they had been mining; he sent ore samples to Quebec, but nothing more was done. Various sightings followed at long intervals. In 1850 a lumberman in the area took out ten tons of ore. By the 1880s, Ontario had an important nickel-mining industry, centring on Sudbury, but events in the previous half-century in the western half of the continent had convinced the world that eastern Canada was not rich in gold and other precious metals, a view that would prove to be ludi-crous. Northern Ontario was neglected because it was undeveloped, and it was undeveloped because it was neglected. In Victoria, South Africa, and the Yukon, railways were built to move ore from the mines, but in northern Ontario mining only became possible once the railway had been put through for other purposes. The railway was the Temiskaming & Northern Ontario (T&NO).

A lumbering company was working the Lake Timiskaming area on a large scale and the fur trade remained important, but there was little other economic activity, or little promise of any, until a number of individuals realized the agricultural potential of what came to be called the Great Clay Belt, a million acres of clay loam amid the otherwise non-arable landscape of the Canadian Shield. On the Quebec side of the lake, an Oblate priest with an entrepreneurial streak formed a colonization scheme. On the Ontario side, C.C. Farr, a former Hudson's Bay Company factor, bought a large tract that he hoped to develop into Little England. In a story reminiscent of John Sutter's New Helvetica, he lived the squire's life, founding his own town, which he called Haileybury after his public school in England, and putting mills into operation. Unlike Sutter, however, he welcomed outside development and indeed since the 1890s had been petitioning the provincial authorities to build a rail line into the area. The promise of new settlement combined with an expansion of lumbering eventually brought a favourable decision. In 1900 surveys began, and the following year the government was sending parties of prospective settlers north for a look. It was in 1903 that two prospectors working as 'timber cruisers' locating timber suitable for railway ties found an outcrop of soft metal, pinkish in colour. Their names were James H. McKinley and Ernest J. Darragh. While both were Canadians, they had had experience in California mines, and they knew they had found something out of the ordinary. They sent samples to a federal government geologist in Ottawa and to a private one at McGill University in Montreal. The former contended that they had found nothing more than bismuth. The latter reported that the rock indicated four thousand ounces of silver to the ton. They chose to believe the latter and staked their claims. Silver was to fulfil the function in Ontario that diamonds had done in South Africa: it defined the gold rush to follow, trained its participants, and provided the financing.

The evidence of past rushes shows that such early discoveries come in clusters and that they are often accidental, though the element of caprice no doubt increases with the telling. The familiar pattern was now about to be repeated. New Liskeard (named after Liskeard in Cornwall) was a tiny settlement on Lake Timiskaming that had sprung up when the rail line emphasized the value of its natural harbour, a much better facility than the one at Haileybury just to the south. In September 1903 Fred

LaRose, a former prospector, was working as a blacksmith for one of the T&NO contractors a few miles south of Haileybury. He had an arrangement with the contractor, Duncan McMartin, that he could prospect in his spare time and that they would split any profits evenly.

Like Edward Bray of South Africa who had discovered gold when throwing a shovel at a Bantu workman, LaRose discovered silver after throwing his smithy's hammer at a fox, or so he always maintained. The hammer broke off a piece of pinkish rock through which threads of some fine mineral were woven. LaRose melted the metal in his forge but still could not identify it. As he recounted later, 'I go to the boss, Duncan McMartin, and I say, "Boss, I have a good thing; come with me." I say, "You give me a good show." He says, "Pull a gun on me if I don't." Then I show him the vein and we stake out two claims, one in his name, another in mine.' Actually, since LaRose was illiterate, he had someone, perhaps McMartin, complete the paperwork, with the result that the name on the records was Frederick Rose.

A hotel keeper in Haileybury who grubstaked local prospectors thought the soft metal was copper in an unusual state. But when the government geologist Willet Miller was called to the scene, he knew at once what it was. Before he could return south, Miller was summoned by another French-Canadian railway worker, Tom Hébert, who reported making a similar discovery near by, one that Miller characterized as 'a text-book vein, it was so perfect.' Miller hurried to Toronto to confer with his superiors. The government announced suspension of all staking for ten miles on either side of the railway's right-of-way until the following spring, a measure designed to allow the appointment of a mining recorder in Haileybury as well as to discourage speculation – though the public, when it heard the news, remained sceptical. Miller returned in the spring accompanied by two prospectors, Richard Anson-Cartwright (died 1958) and William G. Tretheway, a Cornishman who is described in one account as setting off into the bush wearing a starched shirt and a diamond stickpin along with his wellingtons, and carrying a rifle.

Miller bestowed the name Cobalt on the site of LaRose's discovery because it was cobalt bloom that gave the rock its pink colour and because he imagined that the name would help attract attention to the place. Tretheway found another outcrop of the same type, whereupon Miller became more precise, suggesting the name Coniagas, a combination of the symbols for cobalt (Co), nickel (Ni), silver (Ag), and arsenic (As).

When the 1904 mining season sprang to life, all the original discoverers were thus set to meet their various fates. McKinley and Darragh formed a company that was immediately profitable on a vast scale, though it quickly passed into American hands and they retired from the field having made a quick fortune. LaRose was not so lucky. There were legal challenges to his ownership that required a special commission to resolve. During this period, while travelling to his family home in Hull, Quebec, he stopped off in Mattawa and showed some samples to Noah Timmins (1867–1936), who kept a general store and had often grub-staked prospectors, including LaRose. When LaRose had gone, Timmins contacted his brother Henry in Montreal to locate LaRose and make him an offer. After disturbing many a family named LaRose before finding the right one, the brothers set in motion a deal by which the black-smith's portion changed hands for $28,000. It produced many millions' worth of silver. The first fruits of the LaRose Mine (for the illiterate blacksmith at least retained some glory) were bought by William Guggenheim of the American banking family, who had been dealing in the Klondike with spectacular results now that capital had replaced enterprise there.

Tom Hébert quickly sold out as well – to a local syndicate, which in its search for capital to begin mining, turned to a Standard Oil executive. Hébert's claim became the Nipissing Mine, called the Big Nip. As for the Coniagas Mine, Tretheway needed help to fight off a legal challenge of his own and had to barter a half-interest to a Canadian promoter, Colo-nel R.W. Leonard, who was putting together a group of New York State investors for what he therefore called the Buffalo Mine. But Tretheway still managed to reap a fast fortune before selling his remaining interest, in 1906, to a Toronto broker, J.P. Bickell (1884–1951), and an English investor, Colonel Alexander Hay. Before the close of 1904, there were sixteen mines operating in and around Cobalt. There were to be a hun-dred in all before the great Cobalt silver boom ran its peculiar course.

Previous stampedes had been public events (for that is what had made them stampedes); ordinary joes downed tools or jumped counters and hurried off to be a part of the mass adventure; capital made its entrance only at a later stage, either with the importation of full-blown businessmen or the gradual elevation to businessman status of former prospectors, as with Big Alex McDonald and Joe Boyle in the Klondike. Cobalt was different. The silver was sufficiently plentiful and accessible that unaffiliated prospectors could stake mines. Therefore, both pros-pectors and investors were conspicuous from the beginning. The two might even be said to have mixed on terms of mutual admiration.

Martin Nordegg (1868–1948) was a German sent to Canada by a syndicate of his countrymen to investigate the climate for investing. He recalled in his memoirs that he needed only one look at Cobalt in 1906 to confirm that the rush was on:

There came mining men and geologists from all over the world; there came promoters from the east and west of the United States, human hyenas who tried to benefit from the inexperience of the innocents by gambling and betting and selling worthless shares for their hard-earned savings. Business men from Toronto, Montreal, and Ottawa came to convince themselves with their own eyes that the glowing newspaper reports and the pressing tips of the stockbrokers were actually true. Nothing is easier than to make enthusiasts of doubting Thomases by taking them over the ground and showing them prospects and actual ore, and when one of the greatest mines laid bare the famous silver sidewalk, the excitement knew no bounds. Many brokers established offices in Cobalt, and in Toronto new brokers' offices grew like mushrooms. The country became silver-mad. Fortunes were made and lost. How I kept my head cool in this whirlwind is difficult to describe.[1]

Americans and Canadians of all classes rushed to Cobalt, giving it a peak population of eighteen thousand, mostly transient. Arthur Woodcock, the father of George Woodcock the author, scurried there from Winnipeg with a local boxer, Victor McLaglen, but they were unsuccessful as prospectors. Woodcock returned to England before giving Canada another go; McLaglen went to Hollywood and became a famous character actor. Among other nationalities were a large percentage of people with experience at some sort of mining: Russians, Finns, Welshmen, and the ubiquitous Cousin Jacks. The T. Eaton Company, whose mail-order arm had done so well supplying the Klondike trade, now leaped to this occasion as well. Its emphasis was on khaki clothing, which had become popular in the Boer War.

The countryside was low and rocky, with swamps and burnt-out forests. The town itself, while significantly different from American or Australian mining towns, was nonetheless difficult to fit into the ordered pattern of Canadian experience. At first there was no legal townsite and thus no legal means of dealing in property or of regulating building. People built where they chose without reference to straight lines and right angles. Tents abutted shacks of logs and tarpaper, while the homes of the mining managers, who were usually Americans, employed the same materials but on a grand scale. Alfred Arlington

(Double A) Smith lived in a log house complete with a private school-room for his daughters, but brought all the family furniture from Chicago, even the grand piano. Another manager was envied for the culinary skills of his Chinese houseboy. By the time the railway assumed responsibility for surveying the site, streets had to be rerouted around mine shafts and other immovable man-made features.

Safety underground was still primitive. Not until 1910 did hard hats become standard and carbide lamps replace candles. Provincial law, however, did prohibit the sale or consumption of liquor within five miles of a mine, except on a doctor's prescription. The results were two-fold: an influx of underpaid physicians from elsewhere in Ontario – many times the number that a community of such size could absorb in normal circumstances – all offering the same treatment; and a prepon-derance of blind pigs (the term 'speakeasy' was still in the future), which government agents raided from time to time, smashing the bar-rels of whisky with theatrical abandon. Haileybury had no mines within a five-mile radius, so there were busy beverage rooms in all its hotels, whose names illustrate the various influences on culture in the region: the Maple Leaf (native pride), the Attorney (creeping Americanism), and the Vendome (new-money pretentiousness).

Cobalt attracted publicity internationally in a world that had recently reconciled itself to the loss of such anachronisms as silver rushes and gold rushes. Its civic leaders did nothing to discourage the attention and indeed developed their own kind of boosterism based on stock prices, as well as on the promise of personal fulfilment for those actually willing to make the journey. Cobalt's priorities were those of a mining town, however, not those of an ordinary community. It was far behind Skag-way, for instance, in awareness of fire danger, and twenty-five hundred or so were left homeless in a blaze that started in a restaurant in 1909. That same year, one thousand cases of typhoid were reported, for sani-tary conditions were dreadful. Untreated sewage ran down the streets on its way to Lake Cobalt, which soon became useless as a source of drinking water. Entrepreneurs, using wagons in summer and sleds in winter, began touring with water from remoter lakes at five cents a bucket. As early as 1908 the city put in pavement, but it had no high school until 1926. Students were sent to Haileybury, whose school fea-tured instruction in geology, which was the basis of the famous Haileybury School of Mines, opened in 1929 with funding from the mining companies. Cobalt did indulge in a little of the hollow cultural preten-sion of wealthy mining towns everywhere. Its opera house, which seems to have reached its cultural peak with road-company productions of Broadway hits, was located in the Rue de l'Opéra. But then Cobalt was unique among the mining boom towns in that it was never free standing but was always a colony of Toronto. Thus, it did not have to be self-sufficient in its opportunities for conspicuous consumption.

The King Edward Hotel in Toronto, which had opened in 1903, became the seat of power from which Cobalt was run. Desk clerks did not look askance when dirty men in moccasins walked across the marble lobby beneath the cut-glass chandeliers. Bellboys grew accustomed to carrying heavy bags of rock samples. In Cobalt's first year, a northern newspaperman began a publication called the *Northern Miner* but soon had to move it to Toronto where the money was. The *Financial Post* also began publication in 1907 and could not have picked a better moment. It was also in 1907 that the Royal Canadian Mint began operation, tapping the rich new stream of Cobalt silver.[2] That same year the railway began to run a Cobalt Special, which left Toronto every night at nine and was in Cobalt shortly after breakfast. Stephen Leacock wrote of its passage through Orillia with its cargo of romance: 'smiling negros and millionaires with napkins at their chins whirling past in the driving snowstorm.'[3]

On the face of it, the Cobalt boom was more scientifically oriented than previous rushes. Whereas few had paid any attention to William Ogilvie's geological report on the Klondike as published by the Dominion government, the equivalent document by Willet Smith quickly sold eighteen thousand copies. Previous generations of prospectors had been quick to disguise their past or to accept distinctions like forty-niners and sourdough, which suggested amateur status. Now the trend was reversed. Humble prospectors often made brief appearances at mining schools, particularly those in Montreal and Kingston, and it was a social distinction to append the letters PEng to one's name. It might be argued that the need for seeming respectability rose as the level of morality sank. The wrongdoers this time were stock-market manipulators rather than prostitutes and bar-room brawlers.

Speculation in Cobalt stocks was so intense that on one occasion New York called out mounted police to disperse the crowd that had gathered outside the New York Stock Exchange, though only the biggest companies were traded on its board. Stocks of the other companies, and there were many hundred, were bought and sold in Toronto, often to an international clientele. It was during the Cobalt boom in fact that Toronto first acquired its once-universal reputation as a hotbed of crooked dealing. The telephone was now so omnipresent that boiler rooms could use high-pressure sales techniques on sucker lists of small American investors who were immune to the bucket shops, which depended on street trade. In Cobalt as in Toronto, solicitors kept drawerfuls of partially completed incorporation papers, and one of the fixtures of northern Ontario, analogous to the whisky drummers of the old West, was the travelling salesman who dealt in fancy share certificates and other forms of financial engraving and printing.

The first successful gamble in the Cobalt region, one that in fact preceded the discovery of silver, was a syndicate organized by George Taylor, a hardware dealer in New Liskeard who was convinced of the area's mining potential. He and a group of local businessmen and farmers floated 7,761 shares at one dollar each, payable in instalments, of a concern they called Temiskaming & Hudson Bay Mining Company. At their low, the shares traded at eight cents. But at the height of the boom they achieved $300, representing an increase of 39,000 per cent on the original investment. The proprietress of a Cobalt boarding house was left one hundred shares by a prospector in lieu of rent and over the years reaped $150,000 in dividends. One of George Taylor's sons later grubstaked a prospector named Ed Horne (1866–1953), who found what became the Noranda gold mine across the boundary in northwestern Quebec.

Occasional success stories helped fuel the rapid rise in prices, but of course there were catastrophes along the way, the worst of them in 1906 involving Nipissing Mines. Captain Joseph Raphael DeLamar of International Nickel was in control of the company and wished to take it public. He retained the services of a New York promoter, William Boyce Thompson, who in turn interested the Guggenheim family. The Guggenheims sent their own mining engineer to investigate. He was John Hays Hammond (1855–1936), at $250,000 a year the highest-paid salaried employee in America. He travelled to Cobalt in a private railway car with his own valet, chef, and wine steward. The report was favourable, and the Guggenheims subscribed for $2.5 million of the stock and took an option on still more. The stock rose accordingly. When the Guggenheims, sensing trouble, declined to exercise the option, however, the price fell twenty dollars in forty-eight hours. The Guggenheims had recommended the stock to some of their friends and associates and now did the honourable thing and reimbursed them for their losses, a total of $1.4 million. Those who had bought on their own initiative, however, were left to their own devices. But the market recovered, as it always did.

Cobalt was known as a poor man's diggings not only because it produced silver rather than gold but because the silver was being claimed
by prospectors, even by nonprofessional ones. Yet the dynamic of the
silver rush was by no means uncomplicated, like that of the California
gold rush. As late as the 1890s, anyone could prospect in Ontario, after
purchasing a licence and paying a small fee of so much an acre for the
land staked, and you could retain everything you found. When this policy yielded a net loss for the government, it was replaced by a system of
regulations and royalties. In the days of Cobalt, claims were forty acres
but were granted only after it was proved that the land contained a
'valuable mineral in place,' on which royalties were to be paid to the
government as the ore was extracted. But the first stipulation was
ignored or glossed over more often than not, and the royalties went
unpaid. Midway through the rush, the rules were tightened. Fraudulent
or suspicious claims, or ones with royalties in arrears, were reopened for
fresh staking. This had the effect of extending the rush in time as other
factors were extending it geographically.

How to distribute claims on lakebeds was a thorny problem, because
in most cases it was more difficult to meet the 'valuable mineral' stipulation. These claims were also, of course, more expensive to work. One
syndicate formed to work a lakebed was headed by Sir Henry Pellatt,
who later built Toronto's most notable architectural folly, Casa Loma.
But in the case of another lake, the Tory government took the unusual
step of forming a public corporation. In the earliest mines, the ore was
simply 'cobbed' manually – broken apart with hammers and put in
sacks for shipment. Higher levels of investment allowed mine owners to
build concentration plants, at which raw ore went in and silver bullion
came out, along with two main by-products, cobalt (which was sold to
makers of glass and ceramics) and arsenic.

Prospectors who were unsuccessful, which was of course the majority, frequently took jobs as miners in order to feed themselves over the
winter months or until they had enough money to return to the field.
Thus, the role of grubstaker passed from the small businessman to the
corporate owner. But the majority of those working in the mines were
professional miners of one sort or another. Like their South African
counterparts, who hired black workers from a variety of tribes in the
hope that linguistic differences would inhibit collective action, the
Cobalt mine owners made a point of taking on as diverse a group of
nationalities as possible. The strategy was unsuccessful. The Cobalt
Miners' Union was formed as a Canadian local of the Federation of

Western Miners, a Wobbly-directed union based in Denver. Miners in 1906 received from $1.50 to $2.25 for a ten-hour shift. In 1907 the Mine Managers' Association, whose fear of the Wobblies was almost pathological, tried to standardize pay throughout the camps. About 2,250 miners, or three-quarters of the workforce, went out on strike. In the longer view, the issue was not the rate of pay so much as the hours of work. In 1908 the workday was reduced to nine hours, and in 1912 to eight. The prospectors' world was becoming complicated with politics and corporate manoeuvring, and hamstrung by regulations. Yet Cobalt was merely a prelude to what followed discovery of gold in a region called the Porcupine, a hundred miles to the northeast.

The initial circumstances were familiar. Early discoveries were thought unimportant or uneconomical. In 1896 a government geologist, E.M. Burwash, examined gold showings while doing work a dozen miles west of what would later be the town of Matheson. Two years later, another geologist found at least small quantities of gold-bearing quartz 'to be well distributed over the region.' In particular, he singled out the area 'south of the trail to Porcupine Lake as giving promise of reward to the prospector.' An assay of the local quartz in 1901 found $5.20 in gold to the ton, too little to be profitable in so remote a place. But when the T&NO began to build a line northwest from Sudbury, the economic feasibility increased. As at Cobalt, the prospectors – with names such as Bill (Hard Rock) Smith – followed close behind the railway-building crews and fanned out over the rugged countryside. The most storied of them was Reuben D'Aigle, though his renown is as a man who narrowly missed making his fortune.

D'Aigle (1879–1959) was from New Brunswick and had joined the rush to the Klondike at nineteen. Later he crossed over into Alaska, where he was so successful that he required a wheelbarrow to carry his gold dust aboard the steamer when he departed. The news from Cobalt brought him to Ontario. As in the Northwest, he did not linger at the centre but sought outlying territory. In time, he moved west of Porcupine Lake and found a quartz outcrop, staking several claims. Before returning in 1907 with a well-financed expedition, he attended geology classes at Queen's University in Kingston. Yet his expedition had no luck. Although a trail he cut crossed over one of the richest areas in the Porcupine, riches failed to find him a second time. He moved south to Gowganda, where a small, short rush was under way. It was also in 1907 that two Finns, Victor Mansen and Harry Benella, prospecting for a Toronto lawyer and other people, made a rich discovery in the Porcupine that quickly resulted in a mine and a mill. But the facilities were destroyed in a fire the following year. Not until the new rail line was opened in 1909 did the action become heated.

George Bannerman, an amateur prospector whose other activity was driving an ice wagon in Haileybury, was shown some gold-flecked rocks which two trappers, Tom Geddes and Harry Lemon, had found in the country immediately north of Porcupine Lake. Bannerman quickly arranged grubstaking through, of all things, a local church group, and promised to take the trappers in as partners if they would take him to the spot. Then, on the southwestern shore of the lake, came the biggest news of all. John S. Wilson was a Torontonian who claimed to have served in Cuba with Theodore Roosevelt's volunteer cavalry, the Rough Riders, and was now working for the T&NO. Sensing that the circle of discovery was beginning to close in around Porcupine Lake, he put together a rather large expedition with financial backing from William J. (Pop) Edwards, a Chicago plumbing contractor with an interest in the Canadian North.

The records in this particular case are maddeningly contradictory and incomplete, but sometime between June and August 1909, Wilson or some member of his party, possibly one Harry Preston, found a huge dome of gold-bearing quartz. The result was the Dome Mine, incorporated in 1910 and financed by the International Nickel crowd who owned the Big Nip. It paid its first dividend in 1915 and continued paying them for decades. Six men were said to have taken part in the discovery, including one fellow who had recently been taken on as the expedition's cook after happening upon the camp while searching for a lost dog; all six were given some share of the proceeds, though they met a variety of fates. Wilson settled into a company directorship, for example, while Preston died in poverty after being reduced to begging food from the Dome Mine commissary.

The Dome discovery seemed to be proof positive that Ontario prospectors could get rich in gold, not merely in silver. It was the signal, the sign from on high, which the fraternity had been awaiting, and it did much to empty the blind pigs of Cobalt and the beverage rooms of Haileybury. One who made the move was a sometime tool-sharpener, Benjamin Hollinger, who made a modest living buying and selling claims. When a bartender grubstaked him to the extent of $45, he promptly sold a half-interest in his prospects for $55 and then took in a partner, Alex Gillies, who sold part of his for $100. 'I was cutting a discovery post,' Gillies later explained, 'and Benny was pulling the moss off the rocks a few feet away when suddenly he let a roar out of him. At first I thought he was crazy, but when I came over to where he was it was not hard to find the reason. The quartz where he had taken off the moss looked as though someone had dripped a candle along it, but instead of wax it was gold.' They flipped a coin to determine who would stake where. All the surrounding claims eventually proved extremely valuable, but the one Benny Hollinger won on the toss made his name famous. It became the Hollinger Mine, in time the largest gold mine in Canada and one of the longest-lived. The developer was Noah Timmins, who had a large pool of cash at the time not only from the LaRose property but also from a mine that he had recently sold to Bernard Baruch. He got word of Hollinger's find in Montreal and dropped his life of luxury to scurry to the site under terrible conditions, arriving on New Year's Day 1910. The mine was producing before the end of the year, and by 1911 a thirty-stamp mill was in place processing the ore.

A third prospector also made a big discovery in the Porcupine: Sandy McIntyre. His name too has been perpetuated in an important company, but through his own improvidence he never did well financially, though he seemed happy enough with his condition in life. McIntyre was a red-bearded Scot who had changed his name from Alexander Oliphant when fleeing his wife in Glasgow. He appeared in the Porcupine in 1909 as an experienced prospector, but he was travelling with an inexperienced one, a Dutchman named Hans Buttner. When they staked claims north of Benny Hollinger's, Buttner sold out for $10,000, while McIntyre settled for even less. He accepted $300 for a quarter-share and when he had drunk the proceeds sold an eighth share for $25 to an ex-Klondiker lately of Cobalt, Jim Hughes by name. He got $50,000 for what remained, but even that was a pittance for what became known in time as McIntyre Porcupine Mines, a company that eventually granted him a pension so that he would not die in poverty despite his best efforts. When Hughes kindly hired him to prospect around Kirkland Lake, on the railway above New Liskeard, he made an important discovery and in payment was given 150,000 shares in the Teck Hughes Mine. But he sold them for a paltry $4,500. McIntyre was the classic prospector for whom the search was more important than the rewards, even though the dream of riches was what kept him going.

The pattern of disintegration so visible in Cobalt, with its labour strife and dubious financial dealings, was repeated in the Porcupine to the further detriment of the gold rush ideal. The three towns that sprang up around Porcupine Lake were shaped by a series of disastrous fires. In

May 1911 a blaze destroyed the surface plant at Hollinger No. 1, the mine at Benny Hollinger's original discovery site. In July flames obliterated South Porcupine, with its pool rooms and boarding houses, mostly of frame or logs, and the neighbouring community of Pottsville. Only Golden City to the east, the most orderly of the three because it had been laid out by the provincial government, managed to escape devastation. At the Dome Mine, fifty-seven people saved themselves by standing neck-deep in the company reservoir. At the freight yard, a boxcar full of dynamite blew up with a terrible report. Thanks to the T&NO, whose first train had arrived only a few days earlier, the injured could be moved south to hospital. Seasonal fires were raging throughout a huge area; there were 73 known deaths, and it is thought that about 125 other people – prospectors scattered in the bush – also were killed.

At least half a dozen mines were knocked out of operation, but all were rebuilt, and in one case the results were dramatic. In reactivating Hollinger, Noah Timmins not only erected a 40-stamp mill capable of handling 300 tons of ore a day, but he took the opportunity to create an entire town, which he modestly called Timmins. He envisioned this planned community full of middle-class housing for his employees as reaching a population of five thousand, but in time it surpassed that figure by a multiple of nine or ten. He put up the first public building, the Goldfields Hotel, and auctioned off building lots for between $500 and $1,500.

McIntyre Porcupine Mines suffered the least loss in the fire but faced other difficulties. A number of veterans who had come up through rough-and-tumble gold rush towns now had management positions. Richard J. Ennis (died 1951), who built the mill and became general manager, had worked in Colorado, as had one of the partners, Charlie Flynn, who also had Mexican experience. But the company seemed to be in financial distress. Its very first gold bars went from the refinery directly to the lawyers to settle an outstanding debt, and Ennis actually had to hide in the mine to avoid the company's creditors. Attempts to sell more stock were not successful, for investors were hesitant, particularly when the president, Albert Freeman, was indicted on charges that he had used the mails to defraud. It was alleged that he had conspired with Julian Hawthorne, son of the author of *The Scarlet Letter*, and Josiah Quincy, the former mayor of Boston, to separate widows and orphans from their savings. At issue were $600,000 in shares of more or less worthless Cobalt silver mines. It is widely believed that Freeman once convened a directors' meeting in prison after being sentenced to five years (the others received shorter terms). All this was entirely too much when coupled with a strike by two hundred members of the Porcupine Miners' Union (another group affiliated with the Wobblies) over the same issues that had prompted the labour action in Cobalt. The presidency of the company was taken over by J.P. Bickell, the boy wonder of Bay Street who had made his fortune in Chicago speculating in grain. With the assistance of Sir Henry Pellatt, he put the firm right.

The lessons of Cobalt and the Porcupine suggest that despite the prospectors' attempts to adapt to new conditions – by seeking more education, for instance, or acquiring more financial experience – their inability to profit from their discoveries had not grown and may even have decreased. Their best hope had always lain in selling their claims for a quick profit rather than trying to develop mines themselves, and despite a few people's best efforts, this still seemed to be the case. But how soon should they cash in and for how much? The elusive nature of the answers did not slow the process whereby amateur prospectors were squeezed out or forced to join the ranks of the professionals.

That so many of the claimants were Americans is probably due to more than simple colonialism; it doubtless relates to the fact that these were people who felt most acutely the depressurization caused by the closing of the frontiers. The frontiers could be cultural or legal as well as geographical. Frederick W. Schumach (1864–1957) was a Danish-born dispensing chemist who had settled in the United States, where he lived well on the proceeds of a catarrh treatment that was sixty-proof alcohol. When the passage in 1906 of the Pure Food and Drug Act caused sales to drop, he left his base in Ohio for Cobalt and later moved to the Porcupine. He had a mine near the Hollinger property and later sold out to the Timminses.

When developments in the Porcupine had stabilized, attention turned to the area east of the T&NO, and fortunes were made at Kirkland Lake, a body of water named for the daughter of a provincial official and later taken by the city that grew up around the discoveries. William H. Wright (1876–1951), an Englishman who had received land warrants in the Great Clay Belt for his service in the Boer War, found the 'main break,' the ancient fault zone containing gold, in partnership with his brother-in-law Edward Hargreaves (1874–1950). Wright managed to remain in control and emerged a wealthy if unassuming figure of the 1930s, when he financed, for example, the purchase and amalgamation of two Toronto newspapers to create the *Globe and Mail*. In time, even his free land from the government became a profitable mine. But the most conspicuous example of a prospector who made the leap in status to financier was Harry (later Sir Harry) Oakes (1874–1943), a crude and gruff figure who came to a bad end.

Oakes was a graduate engineer from Maine who had searched for gold in Australia, Mexico, and the Philippines. He was in Alaska when news of the Kirkland Lake discoveries reached him. In January 1912 he was in the bar of the Swastika Hotel in the gold town of Swastika, waiting until midnight when a number of old claims were set to expire and be thrown open again for staking. As the deadline neared, he formed a partnership with the Tough brothers, four local contractors who had turned to prospecting. The result was the Tough-Oakes Gold Mine, a big business which Oakes managed to keep alive as though it were a small one, nurturing it without outside support until it could survive unaided. He faced similar obstacles in his next venture, Lake Shore Mines, whose shares once traded for as little as ten cents, but he had faith in Kirkland Lake's wealth, and in time he became one of the richest men in Canada. He later moved to the Bahamas, where he was the victim of a bizarre and grisly ritual murder. The case has long excited sensational writers, some of whom have suggested that the Duke of Windsor, who was colonial governor of the Bahamas at the time, may have been implicated in some fashion.

Robert M. Macdonald was an Australian prospector who followed the course of greatest opportunity, seeking gold above all but settling for whatever other precious substance might be available. He had been a dry-blower in Western Australia and had hunted opals there. He had even been in the pearl trade in the Northern Territory. In his memoirs, he recounts a conversation that involved himself and a number of others like him. It took place in the 1920s, when world wars, automobiles, the wireless, skyscrapers, and modernism were destroying the last traces of the nineteenth-century environment in which such people had flourished.

There were eight of us suffering a period of inactivity, called a holiday, in Townsville. We had got tired of the well-known fields of North Queensland, as townships always grew up around our finds and railways had a habit of throwing out their tentacles after us, and we liked neither towns nor railways. One day, when the mercury in the thermometer seemed to be trying to get out at the top of the tube, the Professor said: 'I'm dead tired of this lazy life, boys; I've read all the latest books and all the last mail's papers from New York, London, Paris and Canton. I'll be forced to read those from Sydney and Melbourne if we stay here much longer.'

'What about going north to the Malay States or Burmah?' suggested Miserable Peter. 'I've never been in those parts.' The tear in his left eye gave his face an expression of sorrow which was misleading.

'No good,' Mac answered. 'There's nothing but tin in the Straits Settlements, and big companies and Chinese have got it all in their hands already. How about Alaska?' 'Not for me, thank you,' smiled Lucky Jack. 'I've had enough of Nome and Dawson City.' 'I propose we have a shot at Rhodesia,' put in Boston Bob. 'That's the only place I don't know.' 'Too civilized,' negatived the Inventor Fellow. 'It's all deep-sinking there, and fancy School of Mines fellows from London run the country.' 'We might go back to New Guinea,' put in one who thought he was a poet and who was known by that title.[4]

Gold was discovered in New Guinea in 1926. Although the whole island was under Australian jurisdiction, only the coastal areas were settled or even explored. Australia, like Britain, seemed to have got an early start on the Great Depression of the 1930s, and the rush that followed the announcement, as well as giving new hope to old prospectors, helped to shorten the dole queues in Sydney and other cities. Many of the participants were shipped in by the big dredging companies. It was only later that three brothers named Leahy decided to go into the interior seeking placer deposits. In 1930, in lush valleys in the rugged central mountains, they found tribes living in Stone Age conditions, some of the one million people who had never seen Europeans before. The Leahys pushed on, leaving a valuable film record, and eventually they did find enough gold to make themselves comfortable. The real wealth, however, took the form of anthropological knowledge.[5]

Another gold rush was taking place in 1926 on the other side of the world, in northern Ontario, above Lake Superior near the border with Manitoba. It might be called the last real gold rush in that it was the last to display the essential traits of California and the Klondike. Unlike New Guinea or for that matter the Porcupine, it was a rush of dreamers as much as of professional prospectors. Corporate planning was now inseparable from prospecting, but at Red Lake, as the place was known, the prospectors – they might once again be called argonauts – were still paramount. They came from many different countries, including ones without gold rush associations. They tolerated great physical hardship to get where they were going. And once they arrived, they found that the magic was not meant to last very long. The Red Lake gold rush in fact lasted only from Christmas 1926 to Christmas 1927.

It can be said of Red Lake as of previous rushes that it could have happened sooner if knowledge of the gold's existence had been the only requirement. In 1872 those ubiquitous and generous Native people who recur so often in such situations helpfully pointed out gold deposits to a government geologist. Thereafter, the fact of the gold's existence was rediscovered or reconfirmed at eight- or ten-year intervals. In 1897 the North Western Ontario Development Corporation, a British firm headed by a minor English noble, found deposits that assayed out at up to $12.50 in gold to the ton, based on gold at $20.00 a troy ounce. The geologist on this expedition was killed when his revolver slipped from its holster, struck a rock, and discharged. Another British syndicate, the Anglo-French Exploration Company, searched in the same area in 1912 but then withdrew until 1926, when the rush was on.

Dr E.L. Bruce of Queen's University, another geologist who would soon enter the story, described Red Lake as 'a beautiful and fascinating place. There seemed to be no end of channels, peninsulas and deep bays. One evening at sunset, the surface of the water took on a deep red glow as far as one could see.' One of the few inhabitants was George Swain (1878–1955?), who trapped and kept a trading post, and dressed like a combination of mountain man and cowboy. For years he had had local silver samples in his possession without knowing quite what they were. When word leaked out and was confirmed by Bruce, there was a flurry of interest and an influx of people from Cobalt and the Porcupine, including even W.G. Tretheway. Several prospectors died in an accident in the wilderness and were not discovered until spring. The two most persistent of the silver seekers were William L. Chennell, a former Klondike stampeder, and his partner Findlay MacCallum, who explored the region in 1922 and each year thereafter (for by that time, gold samples had turned up as well). The *Manitoba Free Press* reported in October 1922 that the mayor of one of the nearest Ontario towns was predicting a gold rush at Red Lake. The story seemed to be so much boosterism by a small-time politician, but it was not.

Findlay MacCallum was at the King Edward Hotel in Toronto when he met another prospector, Lorne Howey, and showed him some ore samples as well as a copy of Bruce's favourable report on Red Lake. Howey took in his brother-in-law as a partner and got him a grubstake. His brother Ray Howey, a prospector working for McIntyre Porcupine Mines, came along as well and brought a fourth man. On 25 June 1925 they made their big find, which resulted in another of those curious bits of dialogue that sound so odd out of context but give such insight into how the people spoke and behaved: 'Looks like we found it!' Lorne

Howey cried excitedly. His brother replied, 'Ain't that what we came for?'

The news spread quickly despite poor communications and the fact that the press had not yet made it public. Jack Hammell (1876–1950), the mining promoter, arrived to stake claims that tied onto those of the Howeys (he later sold them to Dome). Meeting George Swain, he urged him in strong terms to stake a claim himself. It was, he said, a sure means of fortune. But the old trapper stoutly refused. On 25 October the *Ottawa Journal* published the first news story about the Howey discoveries; that was the starter's pistol. Hammell quickly formed a syndicate and raised $50,000.

Wishing to move supplies into Red Lake before freeze-up, Hammell made what proved to be a momentous decision: he chartered aeroplanes, specifically the five Curtis HJ2L flying boats of the Ontario Forestry Service, each capable of carrying six hundred pounds of goods. This was the first important use of aircraft in a gold rush. Bush pilots, a breed who in many ways carried on the ideals of the old prospectors, had been flying in the North since about 1920, and they played an important role at Red Lake, which was only three hundred land miles from Winnipeg, the aviation centre of the West. In New Guinea, the Leahys used aircraft to scout likely placer locations, but this was not feasible in the geological conditions that prevailed in northern Ontario. Yet the planes were invaluable, for they kept Red Lake from the type of starvation that had threatened Dawson. They also provided the modern equivalent of the rich man's route to the Klondike. Those who could afford to do so flew to this last gaudy gold rush, though most people travelled overland or by water or, more commonly, through snow.

Alex Gillies (Benny Hollinger's former partner) had returned to Red Lake with Jack Hammell and was there at Christmas when the onslaught began. 'Here they come!' he told his colleagues. By New Year's, the stream had become a flood. The nearest point on the former Grand Trunk Railway, which was now being managed by Canadian National, was a place called Hudson, Ontario. It consisted of a Hudson's Bay Company store, a small station house, and precious little else. East from Winnipeg, west from Cobalt and Timmins, and north from Toronto a thousand miles away, people suddenly disembarked at this isolated whistle-stop and unloaded great quantities of supplies and equipment, which they transferred to dogsleds for the journey to Red Lake. For the early arrivals, such as a group from Dome Mines, this last leg of the trip might take two weeks, for the party was breaking trail as it went. Latecomers found the going easier and had the added advantage of setting up in recently abandoned camps. But the winter conditions made the journey arduous even for experienced winter travellers, which many of the gold seekers were not.

The *Winnipeg Tribune* began stirring up the story on 2 January, and soon the Toronto papers tried to best one another in their coverage. Frederick Griffin of the *Daily Star* flew there in an open-cockpit plane when the temperature was thirty degrees below zero Fahrenheit. He described 'flying two or three thousand feet above the trail made by men and dogs':

On foot that was a journey of four or five days. We did it in an hour and a half.

Here one found a gold settlement in its first state. As the plane banked over Burntwood Bay half a dozen tents might be seen along the shore, behind them the rock formation, site of the original discovery of hidden gold which had during the winter brought men trudging there by the score to seek a problematical fortune. But where were these men? As the machine glided towards the ice, dark figures came running from their burrows. One found out later that hidden in the bush all along the east shore of the bay were the tents of prospectors, most meagre of shelters, primitive in many cases as the homes of the earliest nomads, in which these treasure seekers lived harsh, monotonous lives while they waited for spring to disclose the claims they had staked.

Red Lake [was a] mining centre in embryo ... an unorganized collection of individual anchorites, without women, without amenities, without amusements, without liquor, without even gambling, as they withstood a siege against cold, isolation, boredom. They were human termites hibernating.[6]

Meanwhile, more were arriving, by no means all of them Canadians and Americans. Naturally, Britons and Australians came. Icelanders and Central Europeans came as well. The Scandinavian nations were particularly well represented. The news spread outwards in concentric circles. One reporter made a game of naming the different nationalities. Publicity begat publicity. The *Ottawa Journal* wrote of Red Lake as 'the magnet drawing prospectors and capital from all parts of the world.' In an age when news stories were often long melodramas, this one seemed particularly ripe. On 5 March a newsreel crew from Fox Movietone shot aerial footage of the rush, and it was being shown in cinemas only a few days later. 'Scoop. First Motion Pictures of Red Lake' was enough to rivet the attention of an audience.

Despite the fact that there were people in the rush with such names as Klondike Bill Woodney, the events of 1898 were now far enough in the past to have an attractively impossible ring. At the height of the excitement, Chaplin's *Gold Rush* was playing at the cinema in Kenora, Ontario, 174 miles away, a town that was vying with Winnipeg for the gold rush trade, much as Victoria and Tacoma had once done with Seattle. For the old, Red Lake was one last opportunity. For the generation of still young people who had gone through the world war, it offered a chance to relive the excitement they had known without the danger they feared. It was a rejuvenating experience for them all.

For everyone who took part there were others who did so vicariously, including of course the faraway investors. In Boston, Joseph P. Kennedy, the future patriarch, was the head of a syndicate with a large stake in the rush. In Toronto, where the securities law was applied lightly, stockbrokers not wishing to wait for listed companies to get in on the action began to back prospectors on their own.

Whether self-employed or not, the people arriving at Hudson found pandemonium. 'Every train was loaded with gold seekers, many suffering from a bout with John Barleycorn,' one witness wrote. 'The din of dogfights, drunks, howling dogs, whistling trains and roaring planes made Hudson the noisiest town on earth. It was hard to find a tree stump without a dog attached to it.'[7] Dogs were as plentiful here as horses had been in the Klondike; the difference was simply that their usefulness was greater and their mortality rate much lower. By March 1926, dogs that might previously have been offered for $25 were bringing $200 each. Dog speculators flourished and so apparently did dog nappers, as pets began disappearing in cities as far afield as Dryden, Kenora, and Sioux Lookout. Practically any canine could be sold, however obvious that it was not a husky. Teams were made up of dogs that had never been in harness; even Airedales were pressed into service, and mutts could be sold under whatever heading seemed marketable to the vendor and plausible to the buyer. There are several recorded instances of dogs giving birth to litters on the trail. Canadian National was charging shippers $80 a ton to send freight from Winnipeg to Hudson in the summer and $250 a ton in the winter. In February it increased its rate for dogs on the same route. They had previously travelled as baggage for only $3.00 but now required payment of $11.00 if caged and $8.50 if not.

Supplies were a problem. Weather and increasing demand caused dangerous shortages near the gold camps, and even at the standard victualling places such as Kenora, prices were dear enough. Lard was a dollar a pound, sugar $35 a bag. Bannock, bacon, beans, and black teas were the staples. Many greenhorns among the eight thousand or so prospectors who made the trip failed to take the dogs' appetites into account. For best performance, each dog needed from three to five pounds of frozen fish each day, though some could make do on cornmeal, and there is one recorded instance of a team subsisting on flour and water. Even a veteran of the bush could face disaster. While crossing a lake, one man saw his whole outfit crash through the ice, but he managed to save the dogs and, by some miracle, to dive to the bottom and find the towline, by which he was able to retrieve both sled and supplies. Another trekker miscalculated his grub by two days and got his dogs through to Red Lake only by letting them chew on strips of a raw moosehide which some hunter had left near the trail. Once safely in camp, a Finnish crew furnished its dogs with a freshly killed wolf and the offal of a deer.

Trappers had been using the trail for at least a century, and it was possible to cross it without a dog team, though sometimes only barely. In one case, two horses were hitched to a toboggan; in another case, three men were hitched to one. Timothy Crowley of Quibell, Ontario, who had been a Klondike stampeder and then took part in the Cobalt, Gowganda, and Porcupine rushes, went to Red Lake with only one horse and a toboggan though he was seventy-two years old. Others simply walked at least part of the way. One man is supposed to have travelled with a wheelbarrow over the hard-packed snow, though that does seem unnecessarily difficult even for someone who relished challenge.

Almost as soon as the rush began, there were demands for a railway to Red Lake from either Winnipeg or Kenora, but although plans were sketched and bills debated, the whole question became mired in party politics and nothing was done. The substitutes were sometimes ingenious if also sometimes impractical. In the spring, vessels ranging from tugs to homemade canoes could travel the many lakes, and local business people built a steamer in Kenora. In time, there was a so-called marine railway, a system of barges and trains. Planes became more frequent and bigger. Snow tractors were used with uneven results; sometimes they had to be pulled out of drifts by horses. A sort of stage line was established using horse-drawn sleighs, with way stations and cook tents every fifteen or twenty miles. Passengers and goods alike were weighed at the start of the journey and charged accordingly. One man who was not dressed warmly enough when he set out stopped midway at a Hudson's Bay Company store and bought heavier clothing for the remainder of the trip; he was assessed an additional three dollars.

The concern for weight was even greater on the number of bush air services that sprang up, as shown by a story about Jack Munro (died 1937), one of the colourful figures of the Red Lake rush. Munro was a Nova Scotian who had toured Canada and the United States as a fighter. In 1904 in Butte, Montana, he had even fought the great Jack Jeffries but had lost on a decision. He then rushed to Cobalt and later became mayor of a nearby gold town, Elk Lake. A war wound had cost him the use of an arm and any hope of fighting again, so here he was in another gold rush. When he climbed into one of Jack Elliot's aeroplanes headed for Red Lake, he carried a piece of angle iron from which he planned to make ice creepers for his boots. Since the rate was one dollar a pound, he was charged an extra dollar for the additional weight.

Elliot was a pioneer pilot with an air service in Hamilton, Ontario, who did not dally in beginning a regularly scheduled service from Hudson to Red Lake. The law required that planes must be overhauled after one hundred hours of flying, so he shipped his two JN4 Canucks by rail. In March he flew the first air mail between Hudson and Red Lake, using a special cancellation that is now considered desirable by collectors. Elliot also took out adverts informing prospectors that they could signal him in a medical emergency by placing green spruce boughs on frozen lakebeds where he could spot them. The need for such help declined once a Red Cross hospital was established in Red Lake later in the year.

Who were the people rushing to Red Lake?

Chris Henrick was part of the Dome party. He came carrying twenty pounds of pork chops (which the dogs ate before the humans could get at them), a typewriter, and office supplies. One of the chief municipal officers of Timmins quit to take part in the rush. So did the head of exploration at Coniagas Mines. Another participant was a lumber-camp cook from Nipigon. Still another was the newly appointed mining recorder rushing to set up shop.

Johnny Reilly, who had participated in the Porcupine and Kirkland Lake rushes and whose most notable eccentricity was rolling naked in the snow each morning before breakfast, joined in as well. Bill Brown (died 1936) came on a mission from the Dominion government to establish a post office. He was rumoured to be the scion of a wealthy English family who had left Oxford without a degree following some indiscretion, no one knew what. He was fluent in Greek, Latin, French, German, Swedish, and Ojibway; the last two were especially useful in Red Lake, with its bushworkers and ex-fishermen from the Scandinavian countries and its Native guides. Brown had been a Hudson's Bay Company factor until he accidentally ignited a powder keg with a cigarette, killing two Natives and almost killing himself.

A Belfaster who was working in a hospital in New Jersey saw a story about Red Lake in a New York paper and said, 'This is for me.' He arrived in an overcoat, oxfords, and galoshes, found Red Lake itself fully staked, and pushed on to one of the outlying areas. Hans Pokum (1906–81) read a similar story in Danzig on the Baltic. When he finally made his small fortune, he threw a party in the royal suite of the Marlborough Hotel in Winnipeg. As the other guests guzzled champagne, he dumped cornflakes on the floor a foot deep to demonstrate to a woman how he had travelled to Red Lake on snowshoes. 'Money was never the dominating factor with me, though I enjoyed spending it,' he once said in typical prospector fashion. 'It was the chase, not the finding, of gold that counted.' The person whose name has forever been linked to Red Lake was George Campbell (1899–1948), a Canadian, whose two brothers had found the so-called staircase of gold that became Dome Porcupine. He discovered the Campbell Red Lake Mine and spent the remainder of his days in the region, allowing people to infer from his rakish and free-spending ways that he was a millionaire, which he was not.

Sandy McIntyre lost no time in going to Red Lake in search of a fulfilling good time, as usual. William Loranger of Ste-Anne-de-la-Pérade, Quebec, invented a sort of early snowmobile using the body of a touring car, but he was not able to put it to the test on the Red Lake trail because the railway mistakenly shipped it to Hudson, Quebec, not Hudson, Ontario. George Punker, the second mate on a Great Lakes freighter, went to Red Lake with an arrangement to send freelance articles to the *Chicago Tribune* and found he could support himself selling claim applications, which he made on his typewriter for a dollar apiece, top copies and carbon copies alike. The printed forms were in short supply since 10,000 claims were staked that year, more than 3,000 of them in a one-month period spanning March and April, and 444 on the single busiest day. There were few charges of claim jumping. It was as though the old prospectors' ethic, the open cabin door with the poke left undisturbed on the crude table, had been given a new life despite what was obviously the corporate underpinning of this last old-fashioned gold rush.

As in previous rushes, there was always money to be made by the people clever enough to offer a needed service. All Ontario, not just its mining camps, was now legally dry, and George Johnston (died 1963) used a dogsled and became the region's first entrepreneur of his kind; a bootlegger operates on land and a rumrunner on water, but there seems to be no term for one who does the same on snow. Robert Alpine MacGregor (1878–1966), a Scot who had prospected at Cobalt, Porcupine, and Kirkland Lake, had a newspaper, the *Red Lake Lookout*, in operation in July 1927.

Those who were not entrepreneurs yet did not choose to mine found that jobs were plentiful. Two Finns, Torvi Pulkinner and Earl Oja, arrived in Hudson and were put to work in the rail yards almost instantly. Until the big mines were operational, there was no large force of paid employees and thus no unions, but small incidents of labour strife could still erupt. A group of Native people hired to widen part of the trail from Hudson went on strike for four dollars a day plus board, an increase of 25 per cent. Some Finns and Swedes, who were chopping wood to feed the boilers of the steam-operated diamond drills that were being brought into the region so laboriously, laid down an ultimatum of their own. They stopped working until they could be assured of an uninterrupted supply of snuff. Management acceded to their demands.

The end of the Red Lake gold rush was as much in keeping with tradition as its beginning had been. As staking became dense in areas close to the lake, prospectors moved farther and farther out, as far as Woman Lake, fifty miles to the east, and then even farther. As a result, discoveries were made that could only be pursued by uprooting. The most important came at the end of August 1927 when Tom Barrow found a golden sidewalk, so called, a seam of gold-bearing quartz eight feet wide and four feet thick, at Narrow, forty miles to the east. Such discoveries siphoned off the energy of Red Lake, and so did the prospect of another winter. By Christmas 1927 only a handful remained, and only the big companies prospered.

L'Envoi

By 1929 there was no reason for prospectors to travel to Red Lake or at any rate no need for them to hurry. Every spot that could be staked had been taken. Both that particular gold rush and the entire gold rush phenomenon were over. The timing is symbolic as well as practical. The Great Depression, with its underlying statements about how interdependent nations had become and how little control individuals have over their own destinies, was the *coup de grâce*.

As it happens, the 1930s were a good time for gold miners but not for gold rushers. The value of gold had been floating in the low twenty-dollar range for a generation. When Franklin Roosevelt fixed the price at thirty-five dollars, it became economical to rework practically any place that had been mined successfully at some point in the past. In all the areas that had witnessed gold rushes – from California, to Australia and New Zealand, to Ontario – there was renewed activity by companies as well as individuals. Every creaking old gold-mining corporation that still had a valid charter seemed to be reinflated, with a corresponding effect in the stock markets' gold indices. Investors were drawn by the escalating share prices as well as by the realization that gold (though now illegal for American citizens to hold and hoard – another edict from Roosevelt) was nonetheless the perfect Depression commodity, as useful in deflationary spirals as in inflationary ones.

The times had the effect of making the professional prospectors more corporate and the amateur ones mere unrelated individuals, unaware of their fellows' existence – anything but part of a crusade. In 1932, for instance, the full-time Canadian prospectors formed an organization designed for self-help, self-regulation, and lobbying. Only a few years earlier, a guild of prospectors would have seemed an oxymoron, like a parliament of anarchists or a convention of hermits. At the same time, in hundreds and perhaps thousands of locations along the tributaries of the Sacramento River and the streams that feed the Yukon, people and families were working over small placer claims and then working over them again. Remoteness from the camera is probably all that kept them from becoming a stock photograph in our historical imagination, like their contemporaries, the Okies or the apple sellers. They even touched off several new discoveries, in Quebec and Manitoba, for example. But they were not gold rushers searching for adventure and fortune. They were victims seeking survival and sustenance. There was no mass movement, no sense of pilgrimage, no romantic exploration.

No one but a company executive (or Robert W. Service) ever became rich by arriving at a gold rush too late. Nor did anyone grow wealthy by being on hand too early, before the potential in the earth's crust suffered itself to be realized. That was the lesson taught by Reuben D'Aigle at Red Lake. At the end of the Second World War, D'Aigle again made premature discoveries, this time of iron ore deposits in Labrador and New Quebec, where in the early 1950s there was a significant influx of different nationalities to build and work the mines. But this was not a mineral rush; it was that peculiarly modern phenomenon, a job rush, a number of whose participants then moved on to the next one, the Kitimat aluminum mine in British Columbia. Such job rushes accompanied all the so-called megaprojects in North America, Australia, and elsewhere during the prosperous 1950s and 1960s. Like the periodic oil booms, they often had one or two features in common with the old gold rushes, especially the temporary economic craziness and the effect of the attraction on many different nationalities and races. But they had none of the underlying faith and optimism, which perhaps could not be reactivated in such an age.

World gold markets have an element of erratic behaviour built into them if only because rises in the price provoke more thorough exploration. In the late 1970s and early 1980s, even the placer deposits already reworked in the Depression were actively exploited once more. But, of course, the easily found placer gold is a quickly exhausted resource, and most of the growth in production comes when the top end of the financial scale is used to sink mines to ever greater depths. South African mines are threatened by the fact that they are now the deepest in the world. Russia's rank is not far below that of South Africa, but its production figures are considered unreliable in the West. Brazil, which until the 1980s had but a modest base in gold mining, later became a top producer, partly as a result of what at first looked very much like an old-fashioned gold rush but was not.

The Serra Pelada in northern Brazil calls to mind the Big Hole in South Africa nine decades earlier. The image of thousands of men and boys climbing wooden ladders with bags of wet earth on their backs could one day be as familiar as that of the thin black line inching up the Chilkoot Pass. It is certainly no less haunting. The extraordinary series of events at Serra Pelada began in February 1980 when a poor Brazilian ranch-hand was bathing in one of the streams that run between the low rolling hills on their way to join the Amazon. He saw something glittering in the water. Like James Marshall of Sutter's Mill, he reached down and plucked out a piece of gold. The results, too, were similar to what took place in California, at least superficially.

Within a month, five thousand *garimpeiros* (miners) were inverting the hill, creating a crater in its place. In time, at least forty thousand persons, as many as if not more than had gone to the Klondike, had hurried into the area, and many times that number spread throughout a large portion of the Amazon basin. They included escaped criminals and ne'er-do-wells, but most were desperately poor peasants simply seeking wages as labourers. In the American manner, the Serra Pelada miners quickly elected a committee to administer rough justice, but the state was never out of sight. By law, gold could be sold only to the government, which maintained tight control over the 6,400 *barrancos*, or claims, each less than seven by ten feet.

The great hole was soon six hundred feet deep and half a mile in width, a pit of red mud with benches cut around the sides. It resembled an ant colony in the amount of activity as well as in its rigid social organization. Hundreds of people struck it rich and became millionaires, *bamburrados*, including one man who in September 1983 found a nugget weighing 137 pounds, one of the largest ever recorded, for which the government gold buyers paid him the equivalent of $1 million. Hundreds of others merely owned one or more claims and engaged starving agricultural workers from the northeast to do the hard labour. In fact, they were called *formigos* – ants. Descending to the bottom of the hole, they filled canvas sacks with forty or fifty pounds of dirt, which they then took to the surface on a series of hundred-foot ladders. A very few might receive a tiny percentage of what they carried; many more were paid a small daily wage; the majority were on piecework, getting the equivalent of fifteen cents a trip and grateful to have it. Nearly everyone working at the mine was Brazilian, thanks to government policy as well as to circumstances.

While Serra Pelada had little of the positive quality of the great gold rushes, it did produce the same sorts of social problem. The amount of mercury used to recover the gold killed nearby rivers. In the slum city that grew up around the pit, there were frequent outbreaks of malaria and meningitis. Prostitution degenerated into actual slavery, with organized rings selling young women from the country for as little as three hundred dollars in gold dust. Many of the *formigos* were little better than slaves themselves.

While events there were moving forward, a different kind of ersatz gold rush was building in northwestern Ontario in an area between Thunder Bay and Sault Ste Marie. Since 1869, prospectors and others had noted the potential for gold in the area around the present town of Marathon. Claims had often been staked and then allowed to lapse amid scepticism that gold found in such volcanic and sedimentary material, rather than in quartz, could ever be mined commercially. It was in this environment that two Timmins prospectors, John Larche (born 1928) and Donald McKinnon (born 1929), waited patiently for several claims to come open in December 1979. They quickly staked what proved to be the centre of the enormously rich Hemlo goldfields, precipitating a staking rush early in 1980 when exploration companies, large and small, sent their prospectors scurrying in; the number of claims filed in the province doubled in a one-year period.[1]

Although Hemlo was indeed discovered, or rediscovered, by two prospectors in checked flannel shirts, what followed was anything but a true gold rush. Instead, an extraordinary wave of speculation in mining stocks gripped Canada and the United States. Much of the activity was focused on the Vancouver Stock Exchange, where people with nicknames such as Murray the Pez and Peter the Rabbit became folk heroes of a sort. The frenzy culminated in a dramatic legal case involving two of the biggest players. In March 1983 the Ontario Supreme Court ordered Lac Minerals Ltd of Toronto to hand over a potential 8.5 million ounces of gold to International Corona Ltd, a company that had begun life operating sandwich shops and had been a mere shell as recently as 1981. The ore was still in the ground at that point, for although a job rush of modest proportions was under way, the initial money making was already completed before any mines were actually operating. This time the fortunes had been made not even by corporate miners but by speculators manipulating blips on computer screens. Such was the final perversion of what gold rushes had represented for so long.

Notes

Introduction: Gold Crusaders

1 B. Traven, *The Treasure of the Sierra Madre* (New York: Knopf, 1935).
2 Captain John Smith, *General Historie of Virginia, New England, and the Summer Islands* (London, 1624).
3 Major William Downie, *Hunting for Gold* (San Francisco: California Publishing Co., 1893).
4 Alexander Del Mar, *A History of Precious Metals from the Earliest Times to the Present* (New York: Cambridge Encyclopedia Co., 1901).
5 Cited in Geoffrey Hindley, *Discover Gold* (London: Orbis Publishing, 1983).
6 See Philippa Pullan, *Frank Harris* (London: Hamish Hamilton, 1975).
7 Mark Twain, *Following the Equator* (Hartford: American Publishing Co., 1897).
8 George Woodcock, *Ravens and Prophets: An Account of Journeys in British Columbia, Alberta, and Southern Alaska* (London: Allan Wingate, 1953).

Chapter 1: The California Delusion

1 For a sense of the popular culture of 1849, see Meade Minnigerode, *The Fabulous Forties, 1840–1850* (New York: Putnam, 1924), and the redoubtable Carl Bode's *The Anatomy of American Popular Culture, 1840–1861* (Berkeley: University of California Press, 1960). The former is impressionistic, the latter expository. Together they make a complete statement.
2 See Richard A. Dwyer and Richard E. Lingenfelter, eds., *Songs of the Gold Rush* (Berkeley: University of California Press, 1964).
3 Mark Twain, *Travels with Mr. Brown* (New York: Knopf, 1940).
4 Van Wyck Brooks, *In the Times of Melville and Whitman* (New York: Dutton, 1947).

5 'Copy of a Summer's Journal to California in 1849,' unpublished manuscript in the collection of Steven Temple of Toronto, whose kindness in lending it is acknowledged most gratefully.

Chapter 2: The Crown and the Southern Cross

1 Robert Hughes, *The Fatal Shore* (London: Collins, 1986).
2 See Angus Murdoch, *Boom Copper: The Story of the First U.S. Mining Boom* (Calumet, Mich.: Ray W. Drier and Louis G. Koepel, 1964).
3 Richard Garrett, *The Search for Prosperity: Emigration from Britain 1815–1930* (London: Wayland Publishers, 1973).
4 Bradford Allen Booth, ed., *The Tireless Traveler: Twenty Letters to the Liverpool Mercury by Anthony Trollope 1875* (Berkeley: University of California Press, 1941).
5 Mason later settled in Canada. Information from 'The Patriarch of Boskung, The Strange History of a Venerable Canadian,' typescript by Watson Kirkconnell, dated 1922 and apparently unpublished, in the collection of David Mason of Toronto.
6 The book was John Sherer's *The Gold-Finder of Australia: How He Went – How He Fared – How He Made His Fortune* (London, 1853). See John N. Molony and T.J. McKenna, 'All That Glistens,' *Labour History* (Australia), May 1977.
7 Geoffrey Blainey, *The Tyranny of Distance* (South Melbourne: Sun Books, 1966).
8 Alan Ward, *A Show of Justice: Racial 'Amalgamation' in Nineteenth-Century New Zealand* (Toronto: University of Toronto Press, 1973).
9 'My Chinese Neighbours. By an Australian Settler,' *Argosy* (London), mid-summer 1866.
10 Mark Twain, *Following the Equator* (Hartford: American Publishing Co., 1897).

Chapter 3: To the Ends of the Empire

1 Quoted in June Callwood, *Portrait of Canada* (Garden City: Doubleday, 1981).
2 See Maria Tippett, *Emily Carr: A Biography* (Toronto: Oxford University Press, 1979).
3 See Graeme Wynn, 'Life on the Goldfields of Victoria: Fifteen Letters,' *Journal of the Royal Australian Historical Society*, March 1979.

Chapter 4: Silver into Gold

1 Dan Elbert Clark, *The West in American History* (New York: Crowell, 1937).

2 See Joan Didion, *Slouching towards Bethlehem* (New York: Farrar, Straus & Giroux, 1968).

3 See Sally Zanjani and Guy Rocha, *The Ignoble Conspiracy: Radicalism on Trial in Nevada* (Reno: University of Nevada Press, 1986).

4 *History of Julian* (Julian: Julian Historical Society, 1969).

5 Rupert Hart-Davies, ed., *The Letters of Oscar Wilde* (London: Rupert Hart-Davies, 1962).

Chapter 5: Titans in South Africa

1 A description of the laws proscribing the rights of Africans, as well as other insights into conditions at the New Rush, can be found in Ruth Fust and Ann Scott's *Olive Schreiner: A Biography* (London: Andre Deutsch, 1980), a life of the South African author and social critic (1855–1920), who wrote *The Story of an African Farm*.

2 Mark Twain, *Following the Equator* (Hartford: American Publishing Co., 1897).

Chapter 6: The Rand and Western Australia

1 Marie Corelli, *Free Opinions Freely Expressed on Certain Phases of Modern Social Life and Conduct* (London: Constable, 1905).

2 Lord Randolph Churchill, *Men, Mines, and Animals in South Africa* (London: Sampson Low, 1892).

3 H.V. Morton, *In Search of South Africa* (London: Methuen, 1948).

4 Lucy Griffith Paré, *The Seeds: The Life Story of a Matriarch* (Montreal: L'Arpent perdue and the Alphonse and Lucy Griffith Paré Foundation, 1984). The author, who also lived in Kalgoorlie, married a French-Canadian mining engineer, immigrated to Canada, and began the Paré business family. Her memoir is an obscure but richly detailed recollection of the Western Australian gold rush as seen through a child's eyes.

Chapter 7: Many Roads to Dawson

1 S.B. Steele, *Forty Years in Canada* (London: Herbert Jenkins, 1915).

2 George Woodcock, *Ravens and Prophets: An Account of Journeys in British Columbia, Alberta, and Southern Alaska* (London: Allan Wingate, 1953).

3 Richard O'Connor, *Jack London: A Biography* (Boston: Little Brown, 1964), gives the most even-handed account of London's Klondike adventure,

though many of his boasts have been debunked in Franklin Walker's *Jack London and the Klondike: The Genesis of an American Writer* (London, 1966).

4 See Pierre Berton, *Hollywood's Canada: The Americanization of Our National Image* (Toronto: McClelland and Stewart, 1975).

5 See Donald Creighton, *Canada's First Century* (Toronto: Macmillan of Canada, 1970).

Chapter 8: Climax and Retreat

1 Susan Chitty, *Gwen John, 1876–1939* (London: Hodder and Stoughton, 1981).

2 See the chapter on mining terms in M.H. Scargill, *A Short History of Canadian English* (Victoria: Sono Nis Press, 1977).

3 Quoted in Richard Harding Davis, *Real Soldiers of Fortune* (New York: Scribner's, 1906).

4 Quoted in Elizabeth Robins, *Raymond and I* (London: Hogarth Press, 1956).

5 Ibid.

6 Arthur Treadwell Walden, *A Dog Puncher on the Yukon* (Montreal: Louis Carrier, 1928).

7 See Stuart N. Lake, *The Life and Times of Wyatt Earp* (Boston: Houghton Mifflin, 1931), and Robert Elman, *Badmen of the West* (Secaucus, N.J.: Castle Books, 1974).

8 See Larry Pointer, *In Search of Butch Cassidy* (Norman: University of Oklahoma Press, 1977).

Chapter 9: Last Stands

1 Martin Nordegg, *The Possibilities of Canada Are Truly Great! Memoirs, 1906–1924*, ed. T.D. Regehr (Toronto: Macmillan of Canada, 1971).

2 See James A. Haxby, *Striking Impressions: The Royal Canadian Mint and Canadian Coinage* (Ottawa: Department of Supply and Services, 1984).

3 Quoted in Robert Bothwell, *A Short History of Ontario* (Edmonton: Hurtig, 1986).

4 Robert M. Macdonald, *Opals and Gold: Wanderings and Work on the Mining and Gem Fields* (London: Fisher Unwin, 1928).

5 See Ian L. Idriess, *Gold-Dust and Ashes: The Romantic Story of the New Guinea Goldfields* (Sydney: Angus and Robertson, 1937), and Bob Connolly and Robin Anderson, *First Contact* (New York: Viking, 1987).

6 Frederick Griffin, *Variety Show: Twenty Years of Watching the News Parade* (Toronto: Macmillan of Canada, 1936).

7 Quoted in Frank Rasky, *Industry in the Wilderness – The People, the Machines: Heritage in Northwestern Ontario* (Toronto: Dundurn Press, 1983).

L'Envoi

1 See Matthew Hart, *Golden Giant: Hemlo and the Rush to Canada's Gold* (Vancouver: Douglas & McIntyre, 1985); Ken Lefolii, *Claims: Adventures in the Gold Trade* (Toronto: Key Porter, 1987); and 'Gold '86: An International Symposium on the Geology of Gold Deposits,' *Proceedings of the Geological Association of Canada*, 1986.

Essay on Sources

It is a telling fact that the bibliographies of many mainstream historical works on gold rushes are full of references to government publications. These reports are almost always statistical in nature, and the statistics, at least the ones about gold production, are seldom of much value, for reasons outlined in the introduction to this book. Historians nonetheless put more faith in such official documents than they do in unsanctioned ones. Yet the subject in this case is one about which most books and booklets, including many of the most useful, are not only local in scope or intent but vernacular in character. They should not be dismissed merely because they are homely.

There are numerous instances in which a memoir or diary by a gold rush participant, though technically a primary source, is of far less value than a work by an amateur historian, a reporter, or a publicist. This statement sounds heretical, but social history about such fringe topics has its own orthodoxy. In any case, it is the local, regional, or narrowly national origin of most gold rush writing that has prevented readers from appreciating the subject in the broad international context. The conspicuous exception is W.P. Morrell's pioneering study, *The Gold Rushes* (London: A. & C. Black, 1940), to which subsequent and slighter works, such as this one and *The Gold Rushes* by Robin May (London: William Luscombe, 1977), owe an obvious debt.

Economists and economic historians have been quicker to see the gold rushes as a single occurrence. A modern example is *Oro y moneda en la historia* (1450–1920) by Pierre Vilar (Barcelona: Ediciones Ariel, 1969), which is available in translation as *A History of Gold and Money, 1450–1920* (London: Verso Editions, 1984). Vilar also attempts the dangerous task of bringing order to gold-production totals and assigning them a dollar figure. It was of course easier to sense the unified nature of the gold rushes once they had come to an end. For that reason alone, there is value in *Gold* (London: The Times, 1933), an anthology by divers experts.

In recent years, books have appeared designed to acquaint the small investor and speculator with the history of gold. *Love of Gold* (New York: Lippincott & Crowell, 1980) by the *New Yorker* writer Emily Hahn is one example. Others are *The Boot of Gold* by Kenneth Blakemore (New York: Stein and Day, 1971) and *Gold: An Illustrated History* by Vincent Buranelli (Maplewood, N.J.: Hammond, 1979). In contrast with the subject they cover, their worth decreases in proportion to their visual attractiveness. In the author's view, the most useful single source on present conditions in gold mining and gold trading is *Gold* by Thomas Patrick Mohide (Toronto: Ontario Minister of Natural Resources, 1981). Its international perspective, detailed figures, and historical context have made it something of a best-seller in the dark world of government publications.

California

The literature of the California gold rush is vast and complex. Sadly, there is no separate bibliography that is adequate, only substantial portions devoted to the subject in the standard bibliographies of Americana and western Americana (the latter is an ever more popular term among librarians, archivists, dealers, and collectors). There is, however, no shortage of significant books that provide an overview. Of those published since the Second World War, the most comprehensive and useful is *Gold Dust: The California Gold Rush and the Forty-niners* by Donald Dale Jackson (New York: Knopf, 1980), a study whose combination of detail and restraint could make it a standard work in the future. But Jackson was by no means the first to attempt such a history or the first to succeed. To take one example, Phil Strong's *Gold in Them Hills: Being an Irreverent History of the Great 1849 Gold Rush* (Garden City: Doubleday, 1957) is a perfectly usable popular account, much better grounded than its subtitle would suggest. So too *The '49ers* by Evelyn Wells and Harry C. Peterson (Garden City: Doubleday, 1949), one of a number of titles published to mark the California gold rush centennial, as readers will note below. The most recent study and one of the most useful is Paula Mitchell Marks's *Precious Dust: The American Gold Rush Era, 1848–1900* (New York: William Morrow, 1994).

Several writers returned to the subject time and again over long careers. These authors, with the most useful of their books, are Archer Butler Hulbert, *Forty-niners: The Chronicle of the California Trail* (Boston: Little, Brown, 1931); Joseph Henry Jackson, *Anybody's Gold: The Story of California's Mining Towns* (New York: Appleton-Century-Croft, 1941); and Oscar Lewis, *Sea Routes to the Gold Fields: The Migration by Water to California in 1849–1852* (New York: Knopf, 1949). A worthwhile appendix to the last of these is *The Nicaragua Route* by David I. Folkman, Jr (Salt Lake City: University of Utah Press, 1972), which argues that too much emphasis has been placed on the Cape Horn and Panama routes. To these should be added *To California by Sea: A Maritime History of the California Gold Rush* by James P. Delgado (Columbia: University of South Carolina Press, 1990).

To one degree or another, all of the above draw their strength from contemporary accounts of the gold rush. These are indeed legion. Diaries continue to appear in print for the first time, while a number of others have been made accessible in facsimile reprint series, as is the case with Daniel B. Woods's *Sixteen Months at the Gold Diggings*, first published by Harper in 1851 but since 1976 one of several such narratives in the Arno Press series, the Far Western Frontier. The preferred edition of the single most important eyewitness account, Bayard Taylor's *Eldorado*, is that edited by Robert Glass Cleland (New York: Knopf, 1949). *Pictures of Gold Rush California*, edited by Milo Milton Quaife (New York: Lakeside Press, 1949), is an anthology of contemporary journals and diaries, and as such continues to have great value. Relatively few foreigners' accounts of the California rush have been published. One of the most intrinsically useful is Jean-Nicolas Perlot's *Gold Seekers: Adventures of a Belgian Argonaut during the Gold Rush Years*, edited by Howard R. Lamar (New Haven: Yale University Press, 1985); see also *Apron Full of Gold: The Letters of Mary Jane Megguier from San Francisco, 1849–1856*, edited by Polly Welts Kaufman (Albuquerque: University of New Mexico Press, 1994).

There are numerous biographies of John Sutter, including *Sutter's Own Story: The Life of General John Augustus Sutter and the History of New Helvetia in the Sacramento Valley* by Erwin G. Guide (New York: Putnam, 1936). Far and away the most reliable is *Sutter: The Man and His Empire* by James Peter Zollinger (New York: Oxford University Press, 1939), though *John Sutter: Rascal and Adventurer* by Marguerite Eyer Wilson (New York: Liveright, 1949) makes fresher use of the subject's papers. *John Sutter and a Wilder West*, a collection of essays edited by Kenneth N. Owens, appeared in 1994 (Lincoln: University of Nebraska Press).

Two trends in modern historiography can easily be shown to apply to work on the California gold rush: the interest in business history and in social history. Until recently, anyone seeking the commercial implications of the gold rush was driven to dubious commissioned books, the best of which would include *Wells Fargo: Advancing the American Frontier* by Edward Hungerford (New York: Random House, 1949). Now there is a new breed typified in *Entrepreneurs of the Old West* by David Dary (New York: Knopf, 1986), Richard H. Peterson's 1977 work *The Bonanza Kings: The Social Origins of Western Mining Entrepeneurs*, now available in a new edition (Norman: University of Oklahoma Press, 1991), and Clark Spence's *Mining Engineers and the Lace-Boat Brigade, 1849–1933* (Moscow, Idaho: University of Idaho Press, 1993). Social history has always been a richer field but has taken new directions of late. As traditionally practised, it led to books such as *The Madams of San Francisco* by Curt Gentry (Garden City: Doubleday, 1964). Among the most useful of these curiosities is *Gold Fever, Being a True Account, Both Horrifying and Hilarious, of the Art of Healing (so-called) during the California Gold Rush* by George W. Groh (New York: Morrow, 1966); the subtitle helps disguise its utility in providing valuable information on the medical aspects of the gold rush. *Gunfighters, Highwaymen and Vigilantes* by Roger D. McGrath (Berkeley: University of California Press, 1987) is a curiously named work of scholarship that attempts to debunk the accepted view of western violence, using the neighbouring mining towns of Bodie, California, and Aurora, Nevada. In recent years, an increasing amount of belated attention has been paid to the role of minorities and women in such events as the gold rush. *Women and Men on the Overland Trail* by John Mack Faragher (New Haven: Yale University Press, 1979) and *Women's Diaries of the Westward Journey*, edited by Lillian Schlissel (New York: Schocken, 1982), are examples. In *Nellie Cashman, Prospector and Trailblazer* (El Paso: Texas Western Press, 1993), Suzann Ledbetler tells the story of a female gold crusader whose life (1850?–1925) covered virtually the entire period and whose venues ranged from Tombstone to Dawson. Especially valuable is *A Mine of Her Own: Women Prospectors in the American West, 1850–1950*, by Sally Zanjani (Lincoln: University of Nebraska Press, 1997).

Australia and New Zealand

By the 1850s, Australia was within the purview of anyone claiming membership in the world of British culture (not quite the same as the commercial or political empire), at least potentially. Yet it was also tantalizingly exotic and seemed – indeed, was – impossibly far away. Guidebooks to the goldfields, or personal journals dressed up in order to highlight the utilitarian nature of the information they contained, seem to have been more important to the Australian rush than similar books had been to the Californian one. There was of course a long tradition of emigrants' guides to other British possessions; with the discovery of gold, the tradition was given new importance. *A Lady's Visit to the Gold Diggings of Australia in 1852–53* by Mrs Ellen Clacy (London, 1853) is an example. There are many others, not excluding Edward Hargraves's own work, *Australia and Its Gold Fields: A Historical Sketch of the Progress of the Australian Colony, from the Earliest Times to the Present Day* (London, 1855), a work as pretentious as its title. Perhaps the single most reliable first-hand account is *Land, Labour and Gold, or Two Years in Victoria with Visits to Sydney and Van Diemen's Land* by William Howitt (London, 1855; available in a 1972 British reprint from Kilmore). The body of the book is a series of letters which Howitt, a Quaker, sent to his family in England. As was also the case with respect to California, the works continued to appear long after the information in them ceased to be current and long after the rush had crested; for example, the two-volume *Life in Victoria, or Victoria in 1853* by William Kelly (London, 1859). Such was the after-market for armchair adventurers.

The quotations in the text from Korzelinski's *Memoirs of Gold-Digging in Australia*, translated and edited by Stanley Rose (St Lucia: University of Queensland Press, 1979), attest to its endearing qualities. It is especially valuable because its author worked at a number of different gold camps and was often witness to important events, though he was not at the Eureka Stockade. Of that event, the most common first-hand account is *The Eureka Stockade* (Melbourne, 1855), which its author, Raffaello Carboni, sold at the site of the battle on the first anniversary of the attack. It continues in print in the Currey O'Neil Australian Classics series. *The Experiences of a Forty-niner during Thirty-four Years' Residence in California and Australia* by Charles D. Ferguson (Cleveland: Williams Publishing, 1888) is a pugnacious account by one of the Americans who took part. There are several accounts in *Gold Fever* (Sydney: Angus & Robertson, 1967), an anthology of gold rush journals and the like, edited by Nancy Keesing; in the more recent edition it is retitled *History of the Australian Gold Rushes by Those Who Were There.* The most valuable book on what took place at Ballarat, however, is *Eureka* (Ringwood: Penguin Books Australia, 1984) by John Molony of the National University in Canberra, whose grandfather was one of the defenders of the stockade and whose assistance the author of the present volume is most eager to acknowledge with gratitude.

The iconography of the Australian gold rushes, especially the Victoria one of the 1850s, is addressed in *Gold and Silver: Photographs of Australian Goldfields from the Holtermann Collection*, edited by Keast Burke (Sydney: William Heinemann Australia, 1973), and in *Bill Peach's Gold* by Bill Peach (Melbourne: Macmillan Australia, 1983). The latter is a popular treatment tied to an Australian Broadcasting Commission documentary.

Perhaps the single most important work on gold in New Zealand is *A History of Goldmining in New Zealand* by J.H.M. Salmon (Wellington: Government Printer, 1963).

British Columbia

In contrast to California gold rush literature, that on the British Columbian rushes, and likewise on the Comstock Lode in Nevada and the goldfields of Colorado, is more likely to fall under the heading local history, with all that implies. There are, of course, exceptions. *The Cariboo Trail* (1920) by Agnes C. Laut, in the thirty-two-volume Chronicles of Canada series, which gathers dust in almost every Canadian library and not a few British ones, should not be overlooked.

But as is the case with California, the most worthwhile evidence comes in contemporary diaries, narratives, and guidebooks. Both Kinahan Cornwallis's
works *The New El Dorado; Or, British Columbia* (1858), and Matthew Macfie's
Vancouver Island and British Columbia (1865), are available in the Far Western
Frontier series from Arno Press. University Microfilms, Ann Arbor, have repub-
lished Margaret McNaughton's 1896 account, *Overland to Cariboo: An Eventful
Journey of Canadian Pioneers to the Gold-Fields of British Columbia in 1862.* It
reminds us of the importance of the Cariboo Road and the overlanders who
used it. *The Cariboo Road* by Mark S. Wade, annotated by Eleanor A. Eastick (Victoria: The Haunted Bookshop, 1979), would appear to be the most thorough and
best-documented work. *Wagon Road North* by Art Downs has a readable text
with a large number of archival photographs of the Cariboo rush. It was first
published in 1960 by Northwest Digest in Quesnel and has frequently been
expanded and revised, most recently in an edition from Foremost Publishing in
Surrey. *Barkerville Days* by Fred Luddit, first published in 1969 and available in a
revised edition (Langley: Mr Paperback, 1980), is of a similar order. So is *Halfway
to the Goldfields: A History of Lillooet* by Lorraine Harris (Vancouver: J.J. Douglas,
1977). These only suggest the possible sources of genuine research amid the tan-
gle of local history publications. *Discover Barkerville – A Gold Rush Adventure: A
Guide to the Town and Its Time* by Richard Thomas Wright (Vancouver: Special
Interest, 1984) has several worthwhile features, including biographies of those
buried in the Barkerville cemetery. Perhaps the most dependable work on the
subject is Gordon R. Elliott's 1958 study *Barkerville, Quesnal and the Cariboo Gold
Rush*, whose most recent edition was published by Douglas & McIntyre (Vancouver, 1975).

Nevada

Here, perhaps the richest single source beyond the standard narrative histories,
a source both detailed and cautious, is *The History of the Comstock Lode 1850–1920*
by Grant H. Smith, a book-length account first published as an issue of the University of Nevada *Bulletin* (1 July 1943) and available in a revised edition from
the Nevada Bureau of Mines and Geology at the university. *The Richest Place on
Earth: The Story of Virginia City, Nevada, and the Heyday of the Comstock Lode*,
edited by Warren Hinckle, the colourful one-eyed San Francisco journalist, in
collaboration with Frederic Hobbs (Boston: Houghton Mifflin, 1978), is a collection of readings and documents in narrative form. But here, too, the inquirer is
often forced to sort through a welter of small nonprofessional booklets and pam-
phlets designed for local audiences or the tourist trade. *Legends of the Comstock
Lode* by Lucius Beebe and Charles Clegg (Carson City: Grahame H. Hardy, 1950)

is typical except in that Beebe was a prolific popular writer in the area, occupying a position somewhat analogous to that of Oscar Lewis in California but farther down on the evolutionary scale. *Lost Bonanzas: Tales of the Legendary Lost Mines of the American West* by Harry Sinclair Drago (New York: Bramhall House, 1966) hints at the existence of an entirely different subcategory of parahistory. *Gold Diggers and Silver Miners: Prostitution and Social Life on the Comstock Lode* by Marion S. Goldman (Ann Arbor: University of Michigan Press, 1981) is an original and refreshing feminist interpretation. *George Wingfield: Owner and Operator of Nevada* by C. Elizabeth Raymond (Reno: University of Nevada Press, 1993) fills an obvious gap. Works on organized labour in the mining camps of the western United States include *James H. Peabody and the Western Federation of Miners* by George G. Suggs, Jr (Norman: University of Oklahoma Press, 1991).

Colorado

Perhaps in part because they were more the playground of millionaire mining men than of rank-and-file prospectors, the Colorado goldfields have left a richer legacy of substantial books, both contemporary and modern. *The Mines of Colorado* by Ovando J. Hollister (1867; reprint, New York: Promontory Press, 1974) is a case in point. Hollister had been editor and publisher of the *Colorado Mining Journal* and provides a mountain of information on sites, companies, deals, and personalities. *Midas of the Rockies: The Story of [Winfield Scott] Stratton and Cripple Creek* by Frank Waters (New York: Covici-Friede, 1937; reprint, Chicago: Swallow Press, 1972) is almost as valuable for the later period. It is perhaps a dangerous generalization, but it seems that Colorado has been much better served by its small local publications than the other regions discussed here. *Cripple Creek: A Quick History of the World's Greatest Gold Camp* by Leland Feitz (Cripple Creek, 1967) and *Ghost Trails to Ghost Towns* by Inez Hunt and Wanetta W. Draper (Chicago: Swallow Press, 1972) are representative of many that are quite useful within their limited scope. Two small works by Brian H. Levine are of a higher order: *Cities of Gold: History and Tales of the Cripple Creek-Victor Mining District* (Colorado Springs: Century One Press, 1981) and *Lowell Thomas' Victor: The Man and the Town* (Colorado Springs: Century One Press, 1982).

South Africa

The standard English-language books on the subject include *Out of the Crucible, Being the Romantic Story of the Witwatersrand Goldfields; and of the Great City which arose in their midst* by Hedley A. Chilvers (London: Cassell, 1929) and *The Romance of the Golden Rand, Being the Romantic Story of the Life and Work of the*

Pioneers of the Witwatersrand – the World's Great Goldfields by William Macdonald (London: Cassell, 1933). Both were written while a number of important participants were alive (but when, to judge by the subtitles, the concept of romance had come into effect). Perhaps the most useful survey is *Gold! Gold! Gold! The Johannesburg Gold Rush* by Eric Rosenthal (New York: Macmillan, 1970). This is a journalistic account by a South African encyclopedist, aimed at an American audience, but it draws deeply from the accumulated knowledge of Rosenthal, his father, and his grandfather. The centennial of the Witwatersrand discoveries produced 'Witwatersrand Gold: The Men, the Money, the Mineralization' by D.A. Pretorius in the *Extended Abstracts* of the Geocongress '86 (Johannesburg: Geological Society of South Africa, 1986).

Stefan Kramer's work *The Last Empire: De Beers, Diamonds, and the World* (New York: Farrar, Straus & Giroux, 1993) tells the racist horror-story of the diamond mines past and present. Curiously, however, the most original modern research concerning the Transvaal goldfields does not concern the oppression of the black and mixed-race miners of that time; virtually all the substantial works in this area are so broad chronologically, or so narrowly focused politically, that they provide no fresh research on this point but merely show how the ill treatment fits into the development of apartheid. The most important contemporary work, in fact, is about the social and financial doings of the randlords. Geoffrey Wheatcroft's admirable study, *The Randlords* (London: Weidenfeld and Nicolson, 1985), can profitably be read in tandem with Jamie Camplin's work, *The Rise of the Plutocrats: Wealth and Power in Edwardian England* (London: Constable, 1978), which deals with such figures as Barnato and Robinson in a broader context. Of more or less general books on South African politics, the one necessary for study of the goldfields is *The Fall of Kruger's Republic* by J.S. Marais (Oxford University Press, 1961), which shows by close argument how the Boer War was the inevitable consequence of the gold rush. As in the other chapters, use has been made of more out-of-the-way vernacular publications on local history. Of special interest is *City Built on Gold* by L.E. Neame (Johannesburg?: Central News Agency, 1959?), because it draws heavily on early newspaper accounts. If one believes that fiction about gold rushes can sometimes be useful by providing a dimension not always found in historical works, one might seek out *The Ridge of Gold* by James Ambrose Brown (New York: St Martin's, 1986).

Western Australia

Geoffrey Blainey's classic narrative, *The Rush That Never Ended: A History of Australian Mining* (Melbourne: Melbourne University Press, 1963), of course deals with much more than the gold rush in Western Australia. But it seems to be the book that best places the developments of the 1890s in a broader context. Blainey is easily the most thoughtful commentator on Australian gold rushes, and he shines most brightly when revealing how personal initiative always gave in to, or was gobbled up by, corporate development. As for memoirs of the events in Western Australia, Sir John Kirwin's *My Life's Adventure* (London: Eyre & Spottiswoode, 1936) is an obvious source. *Coolgardie Gold* by Albert Gaston (London: Arthur H. Stockwell, 1937) and *Under the Coolibah Tree* by Gordon Forbes Young (London: Andrew Melrose, 1953) are less political and more concerned with the life of prospectors, though both show signs of the lengthy retrospect in which they were composed.

The Klondike

Perhaps more books have been written about the Klondike gold rush than about any other. The list includes many recollections of important figures as well as those of minor players whose significance resides more in their typicality or longevity. For decades, however, there was little serious attempt to interpret the whole phenomenon. Until the appearance of Pierre Berton's *Klondike: The Life and Death of the Last Great Gold Rush* (Toronto: McClelland and Stewart, 1958; published in the United States as *Klondike Fever*), the most important work on the subject was a monograph, *Settlement on the Mining Frontier*, by Harold Adams Innis, which comprised one-half of the ninth and final volume of the Frontier of Settlement series (Toronto: Macmillan of Canada, 1936). That Canada's great economic historian, with his fine web of argument and dense style, should overlap in such a way with Canada's most popular writer of narrative history, so devoted to anecdote and colour, is actually rather remarkable. Berton, whose father was a gold rush immigrant to Yukon, was reared in Dawson at a time when vivid evidence of its glory days a generation earlier was not only tangible but omnipresent. It has often been remarked that his lifelong immersion in the subject makes *Klondike* the most satisfying of his books. One can only admire the way he juggled the many venues and the difficult chronology, as well as his success in making the distinction between the U.S. gold rush experience and the Canadian one an accepted part of the public's understanding. It is easy to recreate the impact he must have had after crude books such as Merrill Denison's *Klondike Mike: An Alaskan Odyssey* (New York: Morrow, 1943), a work of amateur hagiography distinguishable from many others mainly by its popularity.

There is no need to repeat here the specifics of Berton's bibliography, but attention should be called to his article 'Gold Rush Writing: The Literature of the Klondike' (*Canadian Literature* 4, Spring 1960), which delves more deeply into the matter. It does seem worth while, however, to mention sources that were outside his avowed purview or that have appeared since 1972, when he published a revised version of his work as *Klondike: The Last Great Gold Rush 1896–1899.*

A recent work on women's experience in the Yukon is *The Real Klondike Kate* by T. Ann Brennan (Fredericton, N.B.: Goose Lane Editions, 1990); its subject, Kathleen Ryan, was one of the first women to cross the Stikine Trail. There are several recent books about the second and corporate phase of gold extraction in the Klondike. The most important is *The Gold Hustlers* by Lewis Green (Anchorage: Alaska Northwest Publishing, 1977), a biography of A.N.C. Treadgold that provides an enormous amount of legal, financial, and technical information about the years of consolidation. There are two modern biographies of Treadgold's nemesis, Joe Boyle. These are *Joe Boyle: King of the Klondike* by William Rodney (Toronto: McGraw-Hill Ryerson, 1974) and *The Sourdough and the Queen: The Many Lives of Klondike Joe Boyle* by Leonard W. Taylor (Toronto: Methuen, 1983). Another comparatively recent biography bearing on the gold rush is *Sam Steele: Lion of the Frontier* by Robert Stewart (Toronto: Doubleday Canada, 1979), a concise but thoughtful work.

Even at this late date, diaries of the Yukon gold rush era continue to surface. One example is *Across Canada to the Klondyke, Being the Journal of a Ten Thousand Mile Tour through the 'Great North West,' July 19th–October 13th, 1900 by 'Colonel D. Streamer,'* edited by Francis Bowles (Toronto: Methuen, 1984). The pseudonymous author was in fact Harry Graham (1874–1936), who became a writer of popular music for London's West End but in 1900 was aide-de-camp to Lord Minto, the governor general of Canada, who made an inspection tour of the Klondike during the period of the boundary dispute. Another diary, kept by E. Hazard Wells, an American newspaper correspondent, has been collated with letters and dispatches as *Magnificence and Misery: A Firsthand Account of the 1897 Klondike Gold Rush*, edited by Randall M. Dodd (Garden City: Doubleday, 1984). *From Duck Lake to Dawson City: The Diary of Eben McAdam's Journey to the Klondike, 1898–1899*, edited by R.G. Moyles (Saskatoon: Western Producer Books, 1977), is of interest because it deals with one of the deadly all-Canadian routes. *Ways Harsh and Wild* by Doris Anderson (Vancouver: J. J. Douglas, 1973) is not a diary but an account, written in the voice of William Walker, Anderson's uncle, describing his and her father's adventures in the Klondike and Alaska. Recent works of interest here include *Gold at Fortymile Creek: Early Days in the Yukon* by Michael Gates (Vancouver: UBC Press, 1994).

The Klondike was the most photographed as well as the most written-about gold rush, and a word about its iconography might not be amiss. *Klondike '98: Hegg's Album of the 1898 Alaska Gold Rush* by Ethel Anderson Becker (Portland: Binfords & Mort, 1949) is an extensively captioned selection of images by

E.A. Hegg, many of them not reproduced elsewhere. *Whiskey and Wild Women: An Amusing Account of the Saloons and Bawds of the Old West* by Cy Martin (New York: Hart Publishing, 1974) is not quite so offensive as the title and subtitle suggest, though it cannot escape being the work of a 'history buff'; its value is that it contains rare Klondike and Alaska photographs. One often finds such plates in local history publications as well. For instance, *Peace River Past: A Canadian Adventure* by David L. Macdonald (Toronto: Venture Press, 1981) contains many photographs of one of the overland routes originating in Edmonton. Once again, however, Berton has produced the most satisfying work: *The Klondike Quest: A Photographic Essay, 1897–1899* (Toronto: McClelland and Stewart, 1983). The January–February 1986 issue of the *Archivist*, published by the National Archives of Canada, is given over to gold rush photography in British Columbia and the Yukon.

Berton uses the writings of the reporter and editor Elmer J. (Stroller) White (1859–1930) as a source and, in the revised edition of *Klondike*, refers to a recent gathering of White's pieces. The book is *'Stroller' White: Tales of a Klondike Newsman*, edited by R.N. De Armond (Vancouver: Mitchell Press, 1969). In the present writer's opinion, however, White's yarns are too exaggerated and his prose style is too florid to be of value as history.

The fiction of the Klondike and Alaska gold rushes is a rich topic. Next to the relevant works of Jack London, Robert W. Service's novel *The Trail of '98* (Toronto: Ryerson, 1910) is the most nearly readable of those published close to the events on which they fed. Whatever its shortcomings, Service's book gives more of what must have been the true feel of the gold rush than any of his collections of doggerel, which are so much better known. There were many rival works, such as *The Gold Trail* by Harold Bindloss (New York: Stokes, 1910). In Jules Verne's novel *The Golden Volcano* (1905), two cousins from Montreal take part in the Klondike rush and are pursued by a villain who is killed in a shower of nuggets when the volcano of the title erupts. The tradition is by no means dead, to judge from *Yukon Gold* by William D. Blankenship (New York: E.P. Dutton, 1977).

Ontario

Large-scale mining in Ontario is far better documented than the province's silver and gold rushes, though books about the former are likely to include some information on the latter. Examples are surveys such as *Two Thousand Miles of Gold from Val d'Or to Yellowknife* by J.B. MacDougall (Toronto: McClelland and Stewart, 1946) and *Free Gold: The Story of Canadian Mining* by Arnold Hoffman (New York: Rinehart, 1947). Also, there is sometimes trustworthy information to be gained from corporate histories, such as *Noranda* by Leslie Roberts (Toronto: Clarke, Irwin, 1956); but though the quality of such commissioned books has risen in recent years, the critical reader must still be vigilant. *Harvest from the Rock: A History of Mining in Ontario* by Philip Smith (Toronto: Macmillan of Canada, 1986) was commissioned by the Ministry of Northern Development and Mines and is an even-handed and heavily anecdotal account, marshalling information from a wide variety of sources.

The Cobalt boom, being so closely tied to northern development and also to fraud and near-fraud in the securities markets, was written about hyperbolically at the time. Works in this style include two curious ones by Anson A. Gard, *The Real Cobalt: The Story of Canada's Marvellous Silver Mining Camp* and *Silverland and Its Stories* (both Toronto: Emerson Press, 1909); these are full of detailed and perhaps reliable information about particular companies and townships. The Toronto *Globe* published a free-standing sixteen-page supplement about Cobalt on 3 October 1908. In recent years, local-history publications have tended to concentrate on the post-individualistic period. Examples include *Cobalt: Year of the Strike, 1919* by B. Hogan (Cobalt: Highway Book Shop, n.d.) and *Yankee Takeover at Cobalt!* by John Patrick Murphy (Cobalt: Highway Book Shop, 1977).

Kirkland Lake has been dealt with adequately in *Three Miles of Gold: The Story of Kirkland Lake* by S.A. Pain (Toronto: Ryerson Press, 1960) and *The Town That Stands on Gold* by Michael Barnes (Cobalt: Highway Book Shop, 1978). Barnes is also the author of *Fortunes in the Ground: Cobalt, Porcupine and Kirkland Lake* (Erin, Ont.: Boston Mills Press, 1986).

At the beginning of this little exercise, mention was made of vernacular history. The perfect example is *The Red Lake Gold Rush* by D.F. Parrott (published by the author, 'Fourth edition 1974'). Parrott is clearly unfamiliar with most of the conventions. Indeed, he frequently sinks into semiliteracy. But his honest desire to set down the record as accurately and coherently as he can, working with primary sources whenever possible and revising and correcting in successive editions, is commendable, even touching. He is an example of an amateur writer striving for something higher without losing his basis in the folk tradition. The subject of gold rushes continues to lend itself to such an approach, as with, for example, *North for Gold: The Red Lake Gold Rush of 1926* by Ruth Weber Russell (North Waterloo: North Waterloo Academic Press, 1987).

Index